Rehearsing the State

RGS-IBG Book Series

For further information about the series and a full list of published and forthcoming titles please visit www.rgsbookseries.com

Published

Forthcoming

Rehearsing the State

The Political Practices of the Tibetan Government-in-Exile

Fiona McConnell

WILEY Blackwell

Library of Congress Cataloging-in-Publication Data applied for.

Hardback 9781118661239
Paperback 9781118661284

A catalogue record for this book is available from the British Library.

Cover image: Palden Gyatso casting his vote on 18 March 2006 © Lobsang Wangyal

Set in 10/12pt Platin by Aptara Inc., New Delhi, India

The information, practices and views in this book are those of the author(s) and do not necessarily reflect the opinion of the Royal Geographical Society (with IBG).

Printed and bound in Malaysia by Vivar Printing Sdn Bhd

1   2016

For Dhasa friends

Contents

List of Figures

Series Editors' Preface

The RGS-IBG Book Series only publishes work of the highest international standing. Its emphasis is on distinctive new developments in human and physical geography, although it is also open to contributions from cognate disciplines whose interests overlap with those of geographers. The Series places strong emphasis on theoretically-informed and empirically-strong texts. Reflecting the vibrant and diverse theoretical and empirical agendas that characterize the contemporary discipline, contributions are expected to inform, challenge and stimulate the reader. Overall, the RGS-IBG Book Series seeks to promote scholarly publications that leave an intellectual mark and change the way readers think about particular issues, methods or theories.

For details on how to submit a proposal please visit:
www.rgsbookseries.com

David Featherstone
University of Glasgow, UK

Tim Allott
University of Manchester, UK

RGS-IBG Book Series Editors

Acknowledgements

Research is never a solo activity and the completion of this book would have been simply impossible without the support, assistance, expertise and friendship of numerous people. First and foremost my thanks go to the staff of the Tibetan Government-in-Exile in Dharamsala and across the exile Tibetan community. Needless to say the willingness of exile government officials to spend time discussing their working lives and future aspirations with me was indispensable to the research that underpins this book. My sincere gratitude also goes to all those from the exile community in Dharamsala, Delhi, Dehra Dun, Bangalore, Bylakuppe, Ladakh and London who generously shared their time, opinions and expertise with me. For reasons of anonymity and space it is impossible to thank each by name, but a number of individuals require special mention. Thupten Samphel, Penpa Tsering, Karma Yeshi and Tenzin Lekshay were invaluable to my early discussions at Gangchen Kyishong, while friends in Dharamsala variously associated with Students for a Free Tibet-India, Friends of Tibet-India, GuChuSum and a number of other NGOs helped to make that hill-station in the Himalayas a home away from home. Particular thanks in that regard to Tenzin Tsundue, Bhuchung D Sonam, Tenchoe, Tenzin Choeying, Karma Sichoe, Damchoe, Tenzin Jigdal, TseNgodup, TseDorj, Tenzin Jamyang, Gelek Namgyal and Lobsang. Tsering Yangkey in particular provided not only generous friendship but also fantastic food, and her early death in 2010 left a hole in many hearts. With this research having begun in 2005, and friendships forged since 2002, I've had the privilege of following the migrations, marriages, children and careers of this amazing group of people, and of gaining insights into the painful personal and community struggles of exile life far from parents, siblings and places that feel like home.

Thanks to Migmar Yangchen's family in Bylakuppe who made me feel so welcome, as did Tenzin A and family in Dharamsala and beyond, and Dala Tsering's family in Choglamsar. In Delhi I spent considerable time with the fantastic team at Empowering the Vision, of which particular thanks to Youdon Aukatsang, Lhakpa Tsering and Tenzin Ngawang. Philippa Carrick, Ricki Hyde-Chambers and Paul

Golding at Tibet Society have been great supporters of my efforts to understand the wider implications of exile Tibetan politics, and my thanks to them for including me in the delegation of British MPs to Delhi and Dharamsala in September 2011. In 2012 I had the pleasure of working with the Foundation for Non-Violent Action in Delhi, where I had fascinating discussions with Rebon Bannerji, Mr Tandon and Chok Tsering. I have also spent many an enjoyable afternoon in lively discussion with students and staff at Jawarhalal Nehru University, where it's been fantastic to see the cohort of Tibetan masters and PhD students grow and graduate. Special thanks to Jigme Yeshi, Lobsang Yangtso and Srikanth Kondapalli. In the wider exile Tibetan community I have learnt a lot from conversations with Ngawang, Dechen Pemba and Tendor, and support received from Tsering Tashi, Thupten Samdup and Chonpel Tsering at the Office of Tibet in London has been particularly important.

My research on exile Tibetan politics has its early roots in my undergraduate dissertation undertaken in the early 2000s at Fitzwilliam College, Cambridge, while the arguments developed in this book have their origins in my doctoral research at the Department of Geography, Queen Mary University of London. That research was funded by an ESRC studentship, with additional support from the University of London's Central Research Fund. I am hugely grateful to Miles Ogborn for his detailed critical feedback and supportive supervision during my PhD. I developed my ideas further during an ESRC postdoctoral fellowship at the School of Geography, Sociology and Politics, Newcastle University in 2010–2011, where the encouragement and friendship of Nina Laurie, Nick Megoran, Alex Jeffrey, Raksha Pande and Alison Williams was invaluable. Many productive conversations were had in the wonderful setting of Tynemouth. Additional research was undertaken during my time as a junior research fellow at Trinity College, Cambridge, where the College's research fund enabled fieldwork in India in 2011 and 2012 and an RGS-IBG Small Research Grant funded research on the diplomatic practices of unrecognised states. Cross-disciplinary conversations with Joya Chatterji, David Washbrook, Bhaskar Vira and Philippa Williams at the Centre for South Asian Studies helped hone many of my arguments, while working with Alice Wilson on questions of legitimacy and exile has been a fantastic experience: she is a scholar of incredible knowledge, diligence and generosity. The writing of this book was completed during my current role at the University of Oxford. Whilst my teaching commitments at Oxford have in many ways delayed the completing of this book, my students at St Catherine's College and those who have taken my course on 'Geopolitics in the margins' have been a wonderful source of feedback and inspiration: thank you.

Some of the arguments made in this book have appeared in articles published in *Political Geography (2009)*, *Geoforum (2012, 2015)*, *Area (2013)*, *Annals of the AAG (2013)*, *Contemporary South Asia (2011)* and *Environment and Planning D: Society and Space (2012)*. I am grateful to the editors and anonymous reviewers for their detailed reading and constructive criticism of my work. Parts of this book have been presented at seminars, workshops and conferences in Cambridge, London, Sheffield, Nottingham, Edinburgh, Singapore, Washington DC, University of Virginia and University of Washington. I am grateful to Phil Howell, Deepta Chopra, Sharath Srinivasan, Nayanika Mathur, David Beckingham, Andrew Barry, Sonali Joshi, Dan Hammett, Alex Vasudevan, Dan Swanton, Rebon Bannerjee, Tashi Rabgey and Craig Jeffrey for their hospitality and critical engagement on these occasions. Valuable feedback has also been received from presentations made at conferences of the Association of American Geographers, Royal Geographical Society (with IBG), International Association for Tibetan Studies, European Association for South Asian Studies, Asian Borderlands Research Network and International Studies Association. With regards the latter, discussions with Nisha Shah have been especially useful in fine-tuning my arguments around territory. Friends and colleagues in Tibet studies in particular have provided invaluable advice, support and encouragement. Special thanks go to Tina Harris, Sara Shneiderman, Mark Turin, Dibyesh Anand, Carole McGranahan, Emily Yeh, Tsering Topgyal, Hildegard Diemberger, Tsering Shakya, Robbie Barnett, Tashi Tsering and Emma Martin. My academic 'home', however, remains in political geography, and support and advice from many colleagues, but especially Jason Dittmer, James Sidaway, Phil Steinberg and Klaus Dodds, has been invaluable.

The current and previous human geography editors of the RGS-IBG Book Series who have overseen the various stages of this publication – Dave Featherstone and Neil Coe – have been fantastic both in terms of offering supportive critique and being reassuringly prompt and encouraging. The same goes for the staff at John Wiley & Sons, Ltd, in particular Jacqueline Scott, whose patience and diligence in keeping things on track have been much appreciated. Anonymous reviewers of the proposal and the manuscript provided much appreciated feedback, critique and suggestions on refining the arguments made. Many thanks also to Ailsa Allen and Edward Oliver for creating the maps that appear as Figures 3.1 and 4.1 respectively, and to *Tibetan Review* for permission to reproduce Losang Gyatso's cartoon (Figure 5.1) and to the Planning Commission for permission to reproduce Figure 6.1. Finally, thanks to Heather and Gordon McConnell for their unquestioning support for a daughter with itchy feet and a wandering mind, Abi Brunswick for her listening ear and humour and, most of all, to Conall Watson for his wisdom, encouragement and love.

Note on Transliteration

The Wylie system of transliteration is popular within Tibetan studies as it remains faithful to original Tibetan spellings. However, as it offers little guide as to the actual pronunciation of Tibetan it can be unwieldy for those unfamiliar with the spoken language. The closest to a standardised phonetic rendering of Tibetan is the Tibetan and Himalayan Library's Simplified Phonetic Transcription system developed by David Germano and Nicolas Tournadre. In order to make Tibetan terms readable in this book I have largely drawn on this system, which is based on the Central Tibetan dialect. Exceptions include terms and titles that have easily recognisable spellings, for example Dalai Lama and Panchen Lama.

With regards to Indian place names, the most commonly used spellings are used here (e.g. Dharamsala rather than Dharamshala).

Chapter One
Introduction

Clinging to the Himalayan mountainside, midway between the Indian market town of Dharamsala and the former British hill-station of McLeod Ganj, is a cluster of low buildings: the headquarters of the Tibetan Government-in-Exile. A bell rings and several dozen people drift out of a two-storey building, the women in traditional Tibetan *chubas*, the men either in dark *chubas* or in maroon monks' robes. They mingle on the veranda, sipping tea, flicking through budget reports and catching up on the political gossip. The second bell sends the Members of Parliament scurrying back to their seats, where the newly installed cable-TV camera is trained on them, broadcasting their every word into Tibetan homes in McLeod Ganj.

Crossing the square in front of the parliament building, I enter the Department of Home, where the mildewed walls and peeling layers of paint are evidence of decades of monsoon rains and attempted repairs. The office of the Additional Secretary is typical of Tibetan Government offices throughout India. There's a photograph of the Dalai Lama, a map of Tibet and one of India, a panorama photograph of the Tibetan capital Lhasa and a promotional calendar from a local Indian printer hopeful for a renewed contract. The Additional Secretary explains to me over Tibetan tea and Indian biscuits that he is currently evaluating agricultural yields from settlements in Orissa, liaising with the Department of Health over a TB awareness programme, and heading a committee to oversee locally elected assemblies. Next door, the Department of Security is screening applications for 'Indian Registration Certificates for Tibetans', with batches of forms ready to be dispatched to the Indian Ministry of Home Affairs in Delhi. At the Department of Information and International Relations the Minister is preparing for a press conference on recent events inside Tibet, and over at the Department of Finance applications are being filed for the 'Green

Rehearsing the State: The Political Practices of the Tibetan Government-in-Exile, First Edition. Fiona McConnell.
© 2016 John Wiley & Sons, Ltd. Published 2016 by John Wiley & Sons, Ltd.

Book', the exile Tibetan identity document that every 'bona fide Tibetan' must hold, but which neither permits the holder to travel, nor offers any legal security.

There is both a familiar mundanity to this scene of bureaucratic busyness, and a sense of 'out of place-ness' (Cresswell 1996). The Tibetans in the monsoon rains of India, Members of Parliament in monastic robes and individuals 'playing' at being state bureaucrats and foreign ministers. This might catch our attention as a somewhat intriguing set-up, but it is also one that is easy to dismiss. No government or state legally recognises the Tibetan Government-in-Exile (TGiE).[1] This institution has no legal jurisdiction over territory in Tibet or in exile, it operates within the sovereign state of India, and it is vilified as a 'separatist political group campaigning against the motherland' by the Chinese Government (Zhu Weiqun cited in Aiming 2011: 8). At first glance, therefore, this government-in-exile appears to be both a powerless pawn in Asian geopolitics and simply another political 'oddity' on the margins of the inter-state system.

But this book is a call to pause a little longer, and to consider both the everyday practicalities and the wider repercussions of what is going on here. In what follows I consider how this exiled and unrecognised government is able to function, what the hopes and goals of its leadership are, and what this case might be able to tell us about the nature of state-like governance more generally. For, though their legal authority is extremely limited, the over 3000 staff of the TGiE are nevertheless attempting to play the state game. And it is the metaphor of play, in the theatrical as well as the ludic sense, that I want to suggest can offer a revealing lens, both for viewing the particularities of this state-like non-state, and for examining the nature of everyday state practices and the pedestal upon which statehood continues to be placed.

Two questions are at the core of this book. First, *how* does the TGiE enact state-like functions from its situation of exile in India and a lack of legal recognition? And, second, *why* is such work put into emulating, or mimicking (cf. Bhabha 1984), this particular form of political organization? In order to address these, the chapters that follow are an exploration of this state-in-waiting; a set of institutions, performances and actors through which the exiled community is practising stateness with the broader aim of one day employing it 'for real' back in the homeland. Or, to frame it in another way, this is a 'rehearsal state', complete with playwrights (the Dalai Lama and, increasingly, the elected Tibetan Prime Minister), designated roles amongst the Tibetan civil service, a dedicated rehearsal space in the exile settlements and audiences ranging from the host state India to the Chinese government and the international community more widely. Through chapters that take aspects of rehearsal in turn – settings, roles, scripts and audiences – I argue that this metaphor speaks to the situation of exile stateness in important ways. First, rehearsal has an inherent but ambiguous temporality: rehearsal is done in anticipation of the 'real' event, but could be indefinite. Second, rehearsal depends on participation, on practice, and on developing expertise, but it also presents the challenge of how to keep people engaged with the broader project. Finally, contingent on belief in a script, in the playwright and in there actually *being* a final performance, rehearsal denotes a deliberate and self-conscious political project.

In tracing out this idea of rehearsing stateness, the book draws on and brings together a series of conceptual debates from political geography, political anthropology and critical international relations. At the core are intersections between post-foundational literature on the everyday state – including the idea of state performances and the relationship between the state and territory – and geographies of temporality. The latter bring into dialogue work on exile and prolonged waiting on the one hand, and ideas around anticipating and imagining futures on the other. Such engagement with theoretical interests around the state, performance, space and time is premised on the assertion that, whilst certainly an unusual political configuration, the exiled Tibetan polity is certainly not unique. As such, the discussion that unfolds in the following pages is set against two key contexts. First is to situate the role and functioning of the TGiE within what is a diverse range of non-state polities, from protectorates and leased territories to de facto states and virtual nations. Second, in focusing on a community that resides *in* but is not *of* South Asia, the following chapters tack between this case and questions of governance, territory and statehood within the Tibet/China/South Asian regional context.

The Case of Exile Tibet

Controversy surrounds the legal, territorial and political status of Tibet. In broad brushstrokes, Chinese authorities maintain that Tibet has been and remains an inalienable part of China's territory (People's Republic of China 1992), whilst Tibetans and their supporters assert that Tibet existed as an independent sovereign state prior to the Chinese occupation in 1949 (DIIR 1996; McCorquodale & Orosz 1994). Tibet is also a nation and territory that has long captured the Western imagination and, to a lesser extent, international media headlines. While the focus of recent attention has been on Chinese government crackdowns on unrest inside Tibet, international 'Free Tibet' protests and the Dalai Lama's meetings with world leaders, this book turns critical attention to a key but often overlooked player in the 'Tibet issue': the Tibetan Government-in-Exile.

In 1950, China's People's Liberation Army (PLA) entered Chamdo in eastern Tibet, and by 1951 had declared Tibet's 'peaceful liberation'. Eight years later, the PLA crushed the Tibetan national uprising in the capital Lhasa, and the Dalai Lama and around 80,000 Tibetans crossed the Himalayas to seek refuge in India, Nepal and Bhutan. Today, the Tibetan diaspora numbers over 128,000, with 74% residing in self-contained settlements and scattered communities in India (Planning Commission 2010).[2] On 29 April 1960 the Dalai Lama re-established the

Tibetan government in the north Indian town of Dharamsala, with the twin task of restoring freedom in Tibet and rehabilitating the Tibetan refugees. Over the following decades the exiled Tibetan community, under the leadership of the Dalai Lama and more recently the democratically elected Tibetan 'Prime Minister', has developed, expanded and institutionalised the TGiE, an exilic political structure that is widely regarded as one of the best organised in the world.

A series of changes have been implemented to restructure the TGiE according to democratic principles and, following reforms in 1991, the government has developed a participatory democracy for the first time in Tibet's history. The Dalai Lama's retirement from political life in March 2011 and his transfer of political authority to the elected exile Tibetan Prime Minister (*Sikyong*) have heralded what is widely seen as a distinct new era of exile Tibetan politics. Meanwhile inside Tibet political tensions have been high since street protests across the plateau in 2008 and over 140 cases of self-immolation.[3] Any resolution to the 'Tibet issue' currently seems a distant dream, especially as the dialogue between Dharamsala and Beijing, begun in 1979, ground to a halt in June 2012 with the resignation of the Dalai Lama's two envoys. With the research for this book conducted between 2006 and 2012, it is against such a backdrop of political change and uncertainty that the following narrative unfolds.

Operating under the constitution-like 'Charter of Tibetans in Exile', the TGiE consists of a legislative parliament with members elected from the diaspora, a judiciary (albeit with limited powers) and an executive body (the *Kashag*) in charge of seven governmental departments. The exile administration's state-like functions include the organization of democratic elections, the provision of health and education services for Tibetans in India and Nepal, a 'voluntary' taxation system and the establishment of quasi embassies abroad. Such claims to legitimacy as the official representative of the Tibetan population are thus made despite being internationally unrecognised, having highly limited judicial and policing powers, and lacking de jure sovereignty over territory in Tibet and in exile. Analysing this situation of legitimacy without legality means going beyond the lenses through which the exile Tibetan case has been viewed to date: those of identity and nationalism (Klieger 1992; Yeh 2007), cultural preservation outside the homeland (Harris 1997; Korom 1997) and socio-cultural adaptation (Goldstein 1978; Subba 1990). Rather, *Rehearsing the State* places the institution of the TGiE centre stage, approaching this polity from a political geography perspective, and focusing on the under-researched issue of its state-like governance strategies within the sovereign space of India. As such, in disrupting conventional binaries of state/non-state, sovereign/non-sovereign and citizen/refugee my aim here is to suggest critical interventions both into how statehood is conceived of spatially and temporally, and into understandings of so-called anomalous polities striving to function in international politics today.

International politics is replete with examples of state-like functions being enacted in non-state-like places. From the Palestine Liberation Organization's state-within-a-state in southern Lebanon in the 1970s, to the functioning of the Irish Free State from 1922 to 1937, and Libyan rebels building a parallel state in Benghazi in spring 2011. In placing the TGiE alongside such examples I am not seeking to compare like with like. Nor does this book engage in a project of categorising or classifying geopolitical 'anomalies'. Not only is a systematic comparison of different polities beyond the remit of this case study-based research (cf. Caspersen 2012; Talmon 1998) but such an exercise arguably does not elucidate broader questions posed here about how and why such 'anomalies' enact distinctly state practices. Where points of comparison *are* made the focus is on parallel practices of 'stateness': the state-like performances, narratives and spaces that are common across communities denied legally recognised statehood. This book is therefore grounded in the perspective that, despite their relatively small population and territorial size, polities such as dependencies, stateless nations and de facto states can provide a valuable window on the nature of international politics. Following the argument that the 'exceptional' has something to tell us about the 'normal', an ethnographic focus on such polities' everyday articulations of statehood exposes the contingent practices that underlie political power in so-called 'conventional' states.

The State as Aspirational: Thinking Across State Spaces, Temporalities and Performances

These broader assertions are considered towards the end of the book; my task here is to sketch out the conceptual framing for the chapters that follow. This book is written from a political geography perspective, by which two key approaches are implied. First, that the relationship between power and space, and how this is articulated in different contexts, is of core concern. In this case, attention is focused on how an ostensibly territory-less polity is able to articulate a degree of sovereign authority and act in state-like ways and, in turn, on how this very 'out of place-ness' facilitates experimentation in governing strategies. Second, writing as a (political) geographer means adopting an integrative approach to theory and methodology. Human geography, in its contemporary guise, is in many ways an outward-looking discipline, and the arguments made in this book draw on, bring into dialogue and seek to speak back to a series of theoretical debates and approaches that have preoccupied scholars in political anthropology, critical strands of international relations and South Asian and Tibetan studies, as well as political geography.

The first of these sets of debates concerns understandings of the state. The state is certainly not the most intuitive conceptual lens through which to view the case of exile Tibet. For a start, this is a situation where the existence of a legally recognised state in the past is disputed and where a state in the future is not only inconceivable under existing political conditions, but is not currently being demanded by the exiled elite. Indeed, if we are to follow legal definitions of the state as a juridical entity of the international system and a government as the exclusive coercive organisation that represents a state (Robinson 2013), then what Tibetans in exile have brought with them and have (re)constructed within India simply counts as neither. However, as the following chapters demonstrate, when viewed through the lens of everyday state practices, the seeming disjuncture of the conventional institution of the state and the case of exile Tibet shed valuable light on the contemporary nature of the state. Such a dialogue needs to be facilitated carefully, and a particular route through state theory has been chosen for this task.

The conventional collapsing of territory, authority and population into a 'single unproblematic actor: the sovereign state' (Biersteker & Weber 1996: 5) has been critiqued from many quarters. Inspired by post-structuralist, feminist and Marxist approaches, a range of scholars have challenged the ontology of the state, drawing on Foucault's ideas on governmentality to posit the state as an emergent ensemble of institutions (Corbridge et al. 2005; Scott 1998), and exploring the plural strategies through which political legitimacy is sought. Speaking to geographical scholarship on the everyday and prosaic state (Gill 2010; Mountz 2010; Painter 2006) this book asserts that, by thereby conceiving of the state not as something concrete there to be observed but rather as a structural effect (Mitchell 1991), then the TGiE does appear to have distinctly state-like attributes and functions. For, while the limitations of applying the concept of the state to this unrecognised exile polity will be woven through the book, central to the argument that follows is the power of 'the state' as an idea and an ideal to aspire to.

Such an assertion in and of itself is arguably not particularly original. Arguments for the 'death' of the state have been roundly challenged in recent years not only by geopolitical events, but also by geographers and political anthropologists, amongst others, who argue for the continued salience of the state model. However, by focusing on a case that is denied legally recognised statehood but nonetheless invests considerable time, effort and resources into enacting a series of state-like functions and practices, this book offers a novel lens onto this endurance of the state model. In addressing the question of why the idea of the state is so appealing to communities who are legally outside of the official state system, I develop the notion of statehood as aspirational. This speaks to, and contributes an original perspective on, debates around the state and affect (Aretxaga 2003; Navaro-Yashin 2002; Stoler 2007). For, whilst the affective qualities of the state have predominantly been understood as dominated by negative experiences (of fear, anxiety, suspicion), the case of the TGiE demonstrates a counter set of affective qualities focused around hope, aspiration and cultural security. Moreover, by focusing on the aims and ambitions of TGiE bureaucrats, the exile settlements as spaces of experimentation in state techniques and the TGiE's constitution and planning documents, the chapters that follow expose how the state as an idea is inherently interwoven with the state as a set of materialities (see Corbridge et al.

2005). I therefore seek to make a conceptual link between the idea of state effects (Mitchell 1991) and state affect.

Underpinning many of these post-foundational approaches to the state is the notion of performance, and I take inspiration from both sociological models of dramaturgy (Goffman 1959; Turner 1974) and work that draws on Judith Butler's (1990) understanding of performativity to explore how the image of a stable state is produced through everyday actions (Campbell 1992; C. Weber 1998). Building on this scholarship, this book adds the idea of rehearsal to other theatrical metaphors used to portray political relations at a range of scales such as mimicry (Bhabha 1984), the stage (Anderson 1996) and improvisation (Jeffrey 2013). By invoking the notion of 'rehearsal' this is therefore an exercise in representing the role and functioning of this exile polity in a way that brings to the fore the provisional and pedagogical dimensions of state practices and performances, and adds weight to assertions that states are in a continual situation of emergence (Jones 2012).

The framework of rehearsal developed in this book also brings to the fore the spatial and temporal contingency of the idea of the state. In terms of the former, this book contributes to debates around the relationship between the state and territory, and the idea of 'state-space' (Brenner et al. 2003) more specifically. As a polity with no jurisdiction over territory either in the homeland (Tibet) or in exile (TGiE operates on land leased from Indian federal states), this case opens up the key question of the extent to which territory is an essential prerequisite to the enactment of state functions and practices. This is a novel approach to examining territorial politics and the relationship between territory and the state as, rather than attending to issues of territorial disputes, conflicts or invasion/occupation, this is a case where territorial limitations have been creatively and innovatively worked around. This will be demonstrated most fully in Chapter 4, where I examine the TGiE's series of tiered government hierarchies, its networks of governmental technologies across India, and the material and symbolic importance placed on the exile settlements.

Finally, the idea of rehearsal developed in this book is a framework that encourages a convergence of thinking on the state and performance with questions of temporality. With the split mandate of dealing with immediate needs in exile and continuing the struggle for a future back in the homeland defining exile communities, cases such as the TGiE have a very particular and acute sense of political temporality. A key feature of exile is being stuck in limbo, of waiting. *Rehearsing the State* engages with a growing body of scholarship examining people's experiences of chronic waiting (e.g. Mains 2007) and, in doing so, it charts the Janus-faced nature of prolonged 'timepass' (Jeffrey 2010). On the one hand the debilitating disillusionment of being 'stuck' in exile, and on the other hand the productive aspects of waiting: of experimentation, preparation and reflection. Central to these more positive attributes is how the temporality of the future shapes the exile present. The exiled Tibetan leadership has, in the past, explicitly stated that the purpose of the time in exile is to experiment with state practices in anticipation of implementing these within a future Tibet (Planning Council 1994). A very particular relationship between the state and temporality is thus being articulated here, and it is one that speaks to a growing body of literature on anticipatory action (Anderson 2010; Collier 2008). In the discussion that follows I set this focus on the forms of imagination, performance and calculation through which futures are made present alongside issues of prolonged waiting and the situation of exile. I thus ask what happens to these anticipatory logics when the time frame is extended indefinitely, and how futures are anticipated and acted upon at the scale of the nation.

Researching a State-That-is-Not-a-State

The TGiE as an institution is, at first glance, a neatly definable entity. It has physical headquarters in Dharamsala with the material attributes of statehood (from a parliament chamber to courtrooms, shelves of official reports and letters headed with the government's emblem), its top bureaucrats have decades of experience, and it has an active media and online presence. Yet, at the same time, this is an often elusive polity. What is particularly challenging in this case is the fact that, especially in its relations with the host state India, the TGiE's authority is rarely openly declared, identified or officially sanctioned (McConnell 2009a). As such, it is methodologically challenging to get a handle on how this polity actually functions on the ground, how it constructs relationships with 'its' people and how it is perceived by external audiences.

Whilst arguably more expedient in this case given the TGiE's lack of recognised status, the challenges of researching the 'everyday state' more generally have been the topic of much debate (Corbridge et al. 2005; Das & Poole 2004). Of particular relevance to the research undertaken here has been the shift within political geography towards ethnographic approaches. In an oft-cited paper on this topic Nick Megoran (2006) argues that engaging with a more sustained focus on agency and on how formal political structures operate and are experienced on the ground requires shifting attention away from the discursive, representational and dramatic aspects of statehood and towards mundane political interactions at the micro-level. A growing body of work has been undertaken in this vein in recent years, notably by anthropologists (Gupta 2012; Hansen & Stepputat 2001, 2005), sociologists (Billig 1995) and geographers (Mountz 2010; Nevins 2002; Secor 2001). Crucially, in enabling a productive 're-peopling' of political geography (Megoran 2006: 625), such attention to the micro-politics of everyday state-making uncovers 'the contingent nature of state power, and the various tensions, fractures and incommensurabilities that characterize state institutions themselves' (Herbert 2000: 554). As such, ethnographic approaches to the state are relevant for starting to address the questions of where and at what level is the state?[4]

Central to this has been a multi-sited approach (Marcus 1995), not only in terms of undertaking research in different Tibetan settlements across India, but also in focusing on a range of different governmental institutions operating at various scales. With regards to the former, qualitative research was undertaken in Dharamsala (Himachal Pradesh), Lugsum Samdupling (Karnataka), Sonamling (Ladakh), Majnu-ka Tilla (Delhi), Clementown and Dekyiling (Uttarakhand) Tibetan settlements between 2006 and 2008, with further interviews conducted in 2010–2012 in Delhi, Dharamsala and London. Discussed in more detail in Chapter 4, these particular settlements were chosen to provide insights into how the TGiE operates and is perceived on the ground in what are physically, socio-economically and institutionally quite different Tibetan spaces within India. In the early stages of this research, an official at the Department of Home stressed the importance of soliciting non-Dharamsala perspectives on the TGiE:

> we don't always see the whole picture. We have nine to five jobs and don't visit the settlements much. We don't get a chance to speak to our people, to see the work of other offices, departments. But it's important to get the big picture, to see we are going in the right direction and what our weaknesses are. For you it is important to speak to our staff on the ground – they will have different opinions, different ways of presenting their thoughts on how this ... on how we work (March 2006)

Over 150 interviews were conducted with TGiE officials, staff of Tibetan non-governmental organizations (NGOs), and a cross-section of exiled Tibetans in these settlements. The latter included both male and female, and monk and lay interviewees, and a range of age groups, occupations, those born in exile, and individuals who had sought refuge in India at different periods in their lives.[5] Indian lawyers, local government officials and journalists were also interviewed, and Indian press archives and parliamentary debates on Tibet consulted to build up a picture of how the TGiE is perceived and engaged with by key actors in the host state. Underpinning this research is an appreciation that the Tibetan community in exile should never be read in the singular. Like any self-defined 'community' this is one with a multitude of experiences, opinions and differences, and that is cross-cut by regional, sectarian and class divisions. A range of perspectives on and experiences of the exile government were sought throughout this research, and the interview quotations in the chapters that follow should thus be read as highly situated accounts. However, at the same time, this is ostensibly a study of the TGiE and its operations in India, and so there is necessarily a focus on members of the India-based diaspora who engage with this polity. As such, it follows in a growing tradition of institutional ethnographies that attend to everyday practices, internal tensions and institutional cultures, and employ an extended case method to draw broad conclusions. Particular inspiration is taken from Yael Navaro-Yashin's (2012) work on statecraft in northern Cyprus, Merje Kuus's (2014) study of European Union bureaucrats and Iver Neumann (2012) on the Norwegian Ministry of Foreign Affairs.

Multiple periods of residential fieldwork in India not only allowed me to conduct follow-up interviews with a number of exile government respondents in Dharamsala and Delhi but also meant I was able to be present at key political moments within the community. These included the run up to and polling day for elections to the Tibetan Parliament-in-Exile and the post of Prime Minister in 2006 and 2011, the Dalai Lama's retirement from political life and statement on his reincarnation in 2011, a number of sessions of the Tibetan Parliament-in-Exile, and commemorations for 'Tibetan National Uprising' on 10 March and 'Democracy Day' on 2 September. I also attended the World Parliamentarians' Convention on Tibet in Edinburgh in 2005, and accompanied a delegation of British MPs to Dharamsala to meet with their Tibetan counterparts in 2011. Taken together, these ritualised performances of statehood, nationalism and diplomacy represent key moments in TGiE's attempts to construct itself as a legitimate polity.

However, in order to get a handle on how this legitimacy is reiterated and sustained, as well as gain an insight into the messiness and contingency of both state politics, and the research process itself (Cook & Crang 1995), attention also needs to focus on mundane, everyday practices of stateness. Following Nigel Thrift's persuasive argument for a shift towards the 'little things'; the '"mundane" objects like files, "mundane" people like clerks and mundane words like "the" – which are crucial to how the geopolitical is translated into being' (Thrift 2000: 380), some of the most revealing insights into the functioning of the TGiE came from unexpected encounters and informal conversations on buses, in cafés and simply from engaging with people's everyday routines in the settlements. Particularly revealing from such informal observations were examples of the inconsistencies of bureaucratic practices, and the everyday hassles of negotiating across Tibetan and Indian administrations: from misspelled names, to paperwork lost between offices and uncooperative local officials. Reflections on such everyday encounters and exchanges, as well as the more ritualised state events noted above, were documented in extensive field journals, excerpts from which appear in the following chapters.

In these offices, institutions and settlements my own positionality as Northern Irish and non-Buddhist meant that I was an obvious outsider. However, my long-term involvement with Tibetan campaigning NGOs in the United Kingdom and in India was a connection that I did have with many respondents in Tibetan settlements scattered across India. The relationship between activism and academia is one that has received productive critical attention in geography over the years (Blomley 1994; Routledge 1996; Ruddick 2004). Whilst certainly not a panacea for the often unequal power dynamics of ethnographic research, especially in the global south, there are a number of ways that engaging in activism alongside ethnographic research approaches can be productive. Activism is about collaboration rather than dependency and, with activism continuing 'back home', it is a long-term commitment rather than an activity confined to being in the field. There are the beginnings of research fatigue within the exile Tibetan community, and I found that organising information evenings for Western tourists, speaking to Tibetan radio about my involvement with the Tibet movement, and collaborating with Tibetan colleagues to organise demonstrations to be potentially productive ways of engaging with and 'giving back' to the exile community. Moreover, having a Tibet activism 'track record' meant that I was an accepted part of an already known community, which facilitated my access to individuals and institutions and the establishment of trust. I am acutely aware that these connections were not always reciprocal nor that collaboration meant that differences were erased but, rather, engagements with Tibet activism was a way of starting to negotiate and familiarise my relations in the field. Finally, whilst I was open about my activist work, there were occasions when my role as activist and that as researcher became 'messy' on the ground. Although *injis* (Westerners) engaging in the wider Tibetan movement is actively encouraged, I had to tread carefully when it came to domestic exile politics. A phrase several TGiE officials used to describe how they negotiated involvement in pro-independence activism with their job for an institution that was advocating for a different future for Tibet resonated with my own experiences. They spoke of having 'two hats'. Deciding which hat to put on when and where was a revealing aspect of my fieldwork, sharpening my awareness of the nature of exile politics, and my own positionality.

Narrating the Rehearsal of Stateness

The argument that develops in the following chapters is not an exhaustive history or indeed a comprehensive overview of the TGiE – for texts that are closer to this see those by the Office of the Dalai Lama (1969) and Stephanie Roemer (2008). Rather, what follows is a series of sustained snapshots of how this institution functions on the ground, how it is received there and how it is presented to the outside world. Emphasis is placed on the period during which fieldwork was undertaken (2006–12), with earlier periods discussed through secondary sources and respondents' recollections. These snapshots are unavoidably selective and are shaped by the networks I had access to, but the aim is to provide a 'feel' for this institution – where it has come from, the challenges that it faces and the directions it might be heading in. Through tacking back and forth between TGiE structures, programmes and functions a broad narrative is built up around the nature of exile politics, the temporalities of statecraft and the idea of the rehearsal state.

The following chapter sets out the theoretical contexts and approaches that underpin this study and sketches out the latter's contribution to thinking on the state and the role of unrecognised polities. Attention focuses on the questions of *how* such polities operate and *why* they have adopted specifically state-like discourses and practices. The chapter positions the argument developed in this book in respect to literature in critical international relations (IR) and critical geopolitics on the 'conceptual unbundling' of sovereignty, territory and statehood, and post-foundational approaches to the state. The case is made for focusing on and developing the concepts of stateness and statecraft and attending to debates around the relationship between the state and territory. Such approaches to the state are then discussed in relation to literature on the temporality of exile and ideas of preparing for unknown futures. Finally, these bodies of scholarship are brought together through the notions of performance and rehearsal.

In relating the concepts and debates set out in Chapter 2 to the case of Tibet, Chapter 3 engages in a series of historical and regional scene-setting. It begins by introducing the characteristics of the Tibetan polity from the founding of the Tibetan Empire in the seventh century, through to the increasingly consolidated nature of Tibetan stateness at the end of the nineteenth century. Attention then turns from internal Tibetan practices of statecraft to contested narratives of and claims to statehood, territory and political authority that underpin conflicting Tibetan and Chinese sovereignty claims. Along the way the nature of Tibet's relations with its neighbours is discussed, as is recent scholarship in Tibet studies that seeks to offer alternative conceptualisations to somewhat polarised Tibetan and Chinese accounts of this region's history. The period of de facto statehood (1911–1949) is discussed in some detail, as is the backstory of the flight of the Dalai Lama and establishment of the TGiE in 1959. In introducing the exile 'cast and plot', the chapter ends by outlining the establishment and institutionalisation of the exile Tibetan government since 1960.

Constituting four approaches to rehearsing stateness, Chapters 4–7 explore key aspects of TGiE's state-like functioning: *rehearsal spaces* in terms of the exile settlements in India; the various *roles* adopted and prescribed within the exile community; *scripts* developed for the practising of statecraft; and the part played by *audiences* of these performances. It is important to note here how I am approaching the heuristic device of 'rehearsal'. The use of metaphors in social science writing brings with it twin dangers: that of overextension to the extent that their application is imprecise and obfuscatory (see critique by Jessop et al. 2008: 389), and that of creating too 'neat' a way of looking at the world that simply overlooks key issues. Mindful of these pitfalls, as well as the 'power of metaphor to illuminate the issues with which we work... [and] establish, clarify and analyse connections, comparisons and meaning' (Howitt 1998: 49), my aim here is to adopt a light touch with respect to this metaphor. 'Rehearsal' is perhaps better described here as a lens through which to view the practices of this polity.

This first take at rehearsing Tibetan stateness focuses on the apparent paradox of a stateless nation managing a series of territorialised settlements. I argue in Chapter 4 that it is in these spaces that the exile administration appears most state-like and, in turn, it is the everyday running of the settlements and the important symbolic role that they play, that is central to TGiE's rehearsal of governance in exile. For, while the Indian state does have a presence within these official Tibetan communities established on its territory, it is exile Tibetan stateness that permeates everyday life in the settlements. Not only are the settlements constructed as 'national' spaces in exile where connections to the homeland are fostered, but also they have been designated as productive spaces of experimentation by the TGiE, where the exile government is actively rehearsing different modes of governance, and where imagined futures are sought to be made present.

Shifting from territorialised forms of state(like) power to questions of legitimacy, Chapter 5 draws on Max Weber's tripartite schema of traditional, charismatic and rational-legal authority to focus on the various roles adopted and prescribed within this rehearsal state. The first of these is the charismatic authority of the Dalai Lama and his role as 'playwright' in this rehearsal of statecraft, particularly in instigating and overseeing the democratisation of exile Tibetan politics. Attention then turns to the TGiE's bureaucrats, focusing on both the role that they play in reproducing this state-in-exile, and how the political objectives of the TGiE are naturalised through everyday routines. The construction of rational-legal authority in this exile polity is thus explored through the establishment of bureaucratic hierarchies, set procedures, staff training programmes and the TGiE's organisational culture. However, the productive aspect of waiting demonstrated by this opportunity to experiment, train and seek advice in the art of statecraft is also counterposed with the possibility that we are seeing the TGiE becoming a permanent government-of-exiles.

Chapter 6 builds on the analysis of the TGiE's articulations of power explored in the previous chapters by focusing on the scripting of the exile population itself. Attending to the forms of power that transform individuals into citizens, this chapter examines the TGiE's management of lives and livelihoods in exile and the construction of Tibetan political subjectivities. The idea of scripting is explored both in terms of written copy (the surveys and plans through which the TGiE comes to know its population) and the broader act of scripting: of discursively constructing an ideal population and seeking to regulate individual behaviour to achieve this. Three aspects of the making of Tibetan subjects are explored: the construction of the (exile) Tibetan population as an entity to be managed; the development of state-like rights and responsibilities through a welfare state system and nascent economy; and the discursive and material construction of Tibetan citizenship in exile. As such, drawing on Foucault's understanding of governmentality, I argue that the TGiE's delineation of distinct domains of governance both creates realms that can be acted upon and distinguishes itself as a government in command of a 'political' sphere.

Broadening the scale of analysis, the focus of the final substantive chapter is on how the TGiE positions itself on the international stage: the strategies and discourses it employs and the role that audiences play in these performances of statecraft. Central to this are the enactment of diplomatic practices including the administration of pseudo embassies and the TGiE's engagement with foreign parliamentarians. Through such public performances of diplomacy the intellectuals of Tibetan statecraft are conspicuously promoting their script to audiences they hope can help turn their rehearsal into a reality. The chapter delineates three distinct audiences that the TGiE engages with: the host state India with whom it has a complex and ambiguous relationship; Western states for whom the exile government seeks to reflect ostensibly Western norms of good governance in order to seek legitimacy and support; and the omnipresent but elusive audience of China. In addition to outlining the series of thus far unfruitful Tibetan delegations to Beijing, the possibility that the very state-like-ness of the TGiE proves a significant barrier to engagement with the Chinese authorities is also discussed.

The concluding chapter broadens the discussion beyond this case and sets out wider applications of the idea of rehearsing the state. The notion of emergent stateness is discussed in terms of its utility in exemplifying the partial and processual nature of the state, and in bringing to the fore the quotidian as a site of state formation. More generally, the idea of rehearsing the state both opens up the important questions of how and why stateness endures and can be so attractive to those who do not have it, and offers a revealing insight into the contingency of 'conventional' states that experiment with political practices as they engage in the global political arena. Finally, I argue that the rehearsal of stateness by communities conventionally excluded from the inter-state system offers a valuable glimpse of possible geopolitical futures. In thus seeking to repluralize our understanding of political space, this examination of a dynamic and innovative polity that is daily articulating aspects of state-ness and forging an alternative space for political authority opens up conceptual space for a more 'progressive geopolitics' (Kearns 2008) and a re-evaluation of the political.

The Politics of Researching Exile Tibet

Researching and writing about most topics related to Tibet is inherently political. Writing about the exile Tibetan government is even more so, and, as such, I want to make the politics behind this book clear at this stage. Not only is it incredibly challenging to solicit the political opinions of Tibetans inside Tibet under the situation in China today, but current Sino-Tibetan relations and future political configurations within Tibet are beyond this book's remit.[6] Within the context in which the book does speak – that of exile Tibet – this is neither a pro- nor an anti-TGiE book. As I write there are heated discussions within the exile Tibetan community over what have become increasingly polarised positions on the future of Tibet: that of genuine autonomy within China (the 'Middle Way Approach', or *Umaylam*, the policy adopted and promoted by the Dalai Lama and the TGiE since 1988), and full Tibetan independence (*Rangzen*). Whilst this fraught political debate forms an important context for this book, *Rehearsing the State* is not a commentary on *Umaylam* vs *Rangzen* per se.

Yet this is not, of course, to say that this is a book devoid of politics. As noted above, I come from a position of political solidarity with the Tibetan cause and, though my work for Tibet support organisations is in many ways distinct from this research, many of the underpinning motivations are shared. As such, the research behind this book may not be a classic example of critical praxis (Wakefield 2007) or critical collaboration (Routledge 2003), but it does share the obligation of treading a fine line between unconditional praise and sustained criticism of the TGiE as an institution. Given the situation of exile, there is an acute fragility to this polity. Thus, constructive criticism is welcomed, but critiques that challenge the foundations of the institution are strongly discouraged. The route chosen here is one of documenting and analysing the achievements, challenges and limitations of the TGiE, and disseminating these to audiences beyond those familiar with this case.

Focusing on what the exile community has established and achieved is certainly not intended to detract attention from the hardships faced by Tibetan refugees over the past decades, or to give a false picture of their security in India. Rather, it is driven by a desire to counter, albeit in a small way, the stereotypes of victim-hood, passivity and indeed pacifism that so dominate representations of (exile) Tibetans. The myth of Shangri-La, premised on Tibet as an idyllic yet forbidden land inhabited by peace-loving Tibetans, is a powerful and enduring one. Within it, the Tibetan nation, people and freedom movement are effectively positioned in a series of moral hierarchies: Tibet as utopia, as virtuous, as victim (Hess 2009). This is a set of representations that not only generates unachievable expectations for the lives that Tibetans lead and denies them agency, but also works to silence violent pasts and presents, has internal contradictions, and is contested – and at times resisted – within the Tibetan community. Deconstructing the myth of Shangri-La is, therefore, 'an ongoing, collective project' (McGranahan 2010: 34). My contribution to this project is to put the practices, limitations and achievements of the exile Tibetan government firmly centre stage.

Endnotes

1 I use 'Tibetan Government-in-Exile' throughout this book rather than 'Central Tibetan Administration' (CTA) as the former is how this institution is commonly referred to by external actors and within the community (*Pö shung*, 'Tibetan government'). The adoption of the term 'CTA' in order to avoid 'offending' China or causing political discomfort for India is briefly discussed in Chapter 7.
2 The figure of 128,000 Tibetans in exile is based on the 2009 Tibetan Demographic Survey undertaken by the TGiE's Planning Commission. Estimates within the community put the population at around 150,000.
3 International Campaign for Tibet website (http://www.savetibet.org/resources/fact-sheets/self-immolations-by-tibetans). See also Self-Immolation as Protest in Tibet in

Chapter Two
Rethinking the (Non)State:
Time/Space/Performance

We must hold on to that strangeness, that eeriness ... so we can employ it to defamiliarize ourselves and grasp the political in contexts that present themselves as "normal", whether they be under pseudo- or proper states.
Navaro-Yashin 2003: 120

This book is, in essence, about the state. Its aim is not to formulate a new theory on the state, but it does seek to offer new angles and insights into how we might conceive of both statehood as an ideal, and state practices as rehearsed through repetition, experimentation and modification. In broader contextual and theoretical terms, choosing to write about the state no longer needs defensive justification as it might have done when hyperbole around rapid globalisations bringing about the 'end of geography' and 'death of the state' (O'Brien 1992; Ohmae 1996) was at its peak. This is not to deny that the state as an institution is undergoing significant transformations. Its borders, economies, monopoly over violence and regulation of social life – from law and education to health and security – are being contested and eroded in a diverse range of ways. Migrants and refugees challenge the hermetic boundaries of the nation-state, while guerrilla groups, warlords and transnational organised crime make a mockery of states' control over violence in many parts of the world. Meanwhile, transnational corporations and supranational institutions such as the European Union (EU) and the International Criminal Court encroach on states' abilities to regulate their economies and determine legal jurisdictions. But then the state has arguably always been under pressure. Since its inception at the Peace of Westphalia the modern nation-state has never

Rehearsing the State: The Political Practices of the Tibetan Government-in-Exile, First Edition. Fiona McConnell.
© 2016 John Wiley & Sons, Ltd. Published 2016 by John Wiley & Sons, Ltd.

been the absolute, bounded container of politics that some realist scholars have portrayed it to be.

Yet, at the same time as we increasingly acknowledge the contingency and socially constructed nature of the state, the statist paradigm remains very much with us. Empirically, the state form demonstrates remarkable persistence and adaptability. Conceptually, the idea of the state continues to undergird the practice of international relations and dominate how we think about the relationship between power and space. Unpacking what we mean by the state, how it has achieved such influence over our geographical imagination and the disaggregated nature of how states function on the ground (Gupta 2012) thus continues to be an important academic endeavour. That said, the choice of case study for contributing to such an exercise here appears, certainly at first glance, to be somewhat counterintuitive. Not only is the relationship between 'the state' and the Tibetan region and its exile politics a problematic and contested one (as discussed in the following chapter), but the raison d'être of this book is to examine a classic case of state*less*ness – the Tibetan diaspora – through the lens of state theory. A number of other conceptual frameworks seem considerably more logical for an examination of polities like the exiled Tibetan administration, from debates around the politics of displacement to legal issues facing refugee communities and the nature of contemporary transnational practices. A thought-provoking book on the political geographies of statelessness could no doubt be based on this case study.[1]

Yet, at the same time, the idea of the state is fundamental to this case. The very existence of this exile government in many ways reinforces the salience of the concept of the state both in the imagination and on the ground: it exists because of the actions of an occupying state; its relationship with the host state India determines what Tibetans living there can and cannot do; the denial of recognition by other states limits TGiE's visibility and voice on the international stage; and, until relatively recently, the aspiration of its leaders was to re-establish an independent Tibetan state. Moreover, as I explore in the following chapters, the TGiE's rehearsal of state-like governance in exile projects a particular image of this community to key international audiences and underpins its internal legitimacy.

The case of exile Tibet is far from unique in terms of demonstrating the legitimising effect of state practices and the enduring allure and power of statehood. As political anthropologist Begoña Aretxaga put it:

> The desire for statehood continues to be intense in many parts of the world, in spite, or perhaps because of, the hollowed-out character of the state. Struggles for statehood help to sustain ethnic conflicts, processes of insurgency and counterinsurgency, war economies, international interventions, refugee camps, and torn societies. The commanding power of the state form can partly be understood because the state holds a sort of meta-capital (Bourdieu 1999), its hallowed form commanding an imagery of power and a screen for political desire as well as fear. (2003: 394)

It is the puzzle of this powerful and paradoxical relationship between stateless-ness and statehood that animates this chapter and provides a route map for the broader theoretical frameworks that underpin this book. My approach to this puzzle is twofold. First, I explore how and why the state – and questions of sovereignty, territory and legitimacy that are bound up with it – are key to understanding the existence, functioning and aspirations of governing authorities like the TGiE. Second, I use the lens of polities that seek to play the state game to re-examine state theory and to explore intersections between the state and questions of spatiality, temporality and performance. Literature on the state is vast and theoretically diverse. As such, what follows is not an overview of the genealogy of the state and its theorisation. Rather, viewing debates around the state through the lens of a polity like the TGiE focuses attention on processes and relations rather than categories and labels, and thus enables particular aspects to be brought into sharp relief. These include: the idea and ideal of the state; the relationship between the state and territory; everyday state practices and the notion of 'stateness'; the temporalities of exile; and the notion of rehearsing statecraft.

The State of Statelessness

Non-state polities have always existed alongside the state since the latter was gradually consolidated as a dominant form of political organisation in the decades and centuries following the 1648 Peace of Westphalia.[2] Within Europe the medieval legacy of highly localised polities (city states, duchies, bishoprics, etc.) coexisting alongside overarching religious and imperial entities persisted in a number of locations, while beyond Europe a plethora of chiefdoms, princely states and nested polities were often only partially replaced by the colonial imposition of sovereign statehood. In more recent times a particularly prolific period for the establishment of new polities was the time between the world wars. The aftermath of the chaos of World War I and the collapse of the Russian, Ottoman, Austro-Hungarian and German empires encouraged an unprecedented wave of state creation and provided abundant opportunities for political creativity. The 'experimental states' (Duara 2003; Roslington 2013) that emerged in this period were predominantly ideologically driven and spanned the full spectrum of politics: from communist and fascist statelets to counter-revolutionary polities and micro-states based on nationalist and religious ideologies. As Prasenjit Duara notes in his fascinating account of the Japanese puppet state of Manchukuo, these 'experiments in limited political or electoral representation, nationalist forms, and developmental agendas' (2003: 1) were short-lived entities that declared sovereignty and took on some of the symbolic forms of statehood, but were overwhelmingly unrecognised.

Unfolding alongside the dramatic increase in the number of nation-states since the end of World War II – from 51 members of the United Nations (UN) in 1945 to 193 at present – has been the proliferation of political entities that do not fit the traditional nation-state model. Though varied in their territorial scope, population size and political aims, such polities are consistently the manifestation of intractable tensions between legal principles of self-determination and territorial integrity (McConnell 2009b; Pegg 1998). We can think of League of Nation Mandates, which later became UN Trust Territories (see Anghie 2002), Non Self-Governing Territories such as Western Sahara, and the broader category of dependent territories (Armstrong & Read 2000), all of which resulted from partial or stalled decolonisation. And of course the era of protectorates is by no means behind us as the examples of 'foreign' (read 'Western') administrators, peacekeepers and troops in the likes of Kosovo, Afghanistan and East Timor testify (Mayall & de Oliveira 2011). Then there are cases of shared sovereignty – binational territories (e.g. Brčko district of Bosnia and Herzegovina; see Jeffrey 2006), 'trans-state entities' (e.g. the Indonesia-Malaysia-Singapore Growth Triangle; see Relyea 1998), enclaves (Vinokurov 2007) and leased territories (e.g. Hong Kong 1898–1997; see Strauss 2007) – where the jurisdiction of one state extends over the territory of another.

The end of the Cold War heralded another proliferation of nation-building with over 20 new states emerging from the dissolution of 'parent' states within a few years. But again there is a parallel story. For every case of successful secession there are many where claims to self-determination are thwarted by the resolute upholding of the principle of territorial integrity. These include national liberation or insurgent movements such as the PKK (Kurdistan Workers Party) and LTTE (Liberation Tigers of Tamil Eelam), complex autonomous arrangements and national movements within and across states such as Catalonia and Kurdistan, and de facto states such as Somaliland and Abkhazia (Bahcheli et al. 2004; Kingston & Spears 2004).

As noted in the previous chapter, my intention in this book is not to assemble a comparative framework for so-called 'geopolitical anomalies' (see also Ferguson & Mansbach 1996). The range of historical contexts, legal arrangements and political ideologies listed above questions the utility of imposing order on such diversity, and indeed typologies that have been proposed are often as much about the fuzziness of the boundaries between categories as they are about defining different types of polity (e.g. Caspersen 2012). However, the continued formation, existence and functioning of such non-state polities in and of itself raises important questions that speak directly to political geography's concerns with the relationship between power and space.

Out-of-Place and Out-of-Time: The Spatialities and Temporalities of Exile

It was a conundrum around the relationship between sovereignty and territory that first focused my attention on the TGiE: how does a polity that is internationally unrecognised and lacks jurisdiction over territory both in the homeland and in exile nevertheless appear to articulate forms of state-like sovereign authority? What intrigued me was precisely the TGiE's existence and decades-long functioning inside the state of India. Drawing on Scott Pegg's analytical framework of the birth, life and death of de facto states, it is the 'lifetime' of this exiled government functioning inside a host state that is of particular interest here. This stands in contrast to existing literature on governments-in-exile which focuses overwhelmingly on the 'birth' and 'death' stages of this cycle (see Shain 1991; Vigne 1987). Although for most governments-in-exile their sojourn in exile is short-lived, with their (self)declaration a symbolic strategy to elicit international support without the intention of actually operating as a government whilst in exile (e.g. National Coalition Government of the Union of Burma, 1990–2012), for a small minority this period is protracted. Examples of such governments (or states)-in-exile in recent decades include the TGiE, the Sahrawi Arab Democratic Republic (SADR) established on Algerian territory in 1976 following Moroccan occupation of Western Sahara (Shelley 2004; Wilson forthcoming), and the Palestinian Liberation Organisation (PLO), which has enjoyed observer status at the UN since 1974 (Moshe 1996).

More than simply a political and legal technique (Reisman 1991), these exile polities have an additional and distinctly state-like set of functions compared with more 'conventional' war-time governments-in-exile (Conway & Gotovitch 2001; Dufoix 2002). These functions include the provision of welfare services for a sizeable dependent population, the promotion of nation-building policies, the establishment of democratic institutions, and their operation within (though often not jurisdiction over) defined territories in a host state. Central to these roles is the exile government's relationship with distinct 'constituencies': the diaspora in exile, the population in the homeland and potential supporters within the international community. These polities therefore constitute very particular configurations of sovereignty and territory and make explicit claims to legitimate governance. They are holders of state-potential, builders of social capital and have intriguing parallels to, but lack direct equivalence with, both the categories of governments-in-exile and de facto states. In light of these traits that exceed existing classifications of non-state polities, my aim is not to pin down exactly what kind of political entity the TGiE *is* but rather to consider what it *does*.[3] It is a polity that appears to exist between and across a range of political binaries – statehood/statelessness, sovereign/non-sovereign, refugeehood/citizenry – and this book challenges not only the power-laden construction of such labels and dualisms (see Jones 2009; Moncrieffe & Eyben 2007), but also their conceptual mapping onto each other. In the chapters that follow, in place of such typologies, I develop a critical analysis of the notions of ambiguity and ambivalence in relation both to the emulation of territorialised state practices by stateless polities and to the role that distinctive temporalities of exile play vis-à-vis claims to legitimacy.

Conventionally distinguished from refugees in terms of regarding their separation from the homeland as temporary, exile communities have a very particular and one might say acute sense of political temporality (Rose 2005). The suspended animation of exile and uncertainties regarding the political future therefore define exile polities and populations, and the Tibetan diaspora is no exception. Balancing these different temporalities is challenging, and the split mandate of continuing the struggle for the homeland and dealing with immediate needs in exile creates conflicting responsibilities on both personal and institutional levels.

The limbo of statelessness more generally has been well documented. Polities of indeterminate status – from de facto states, to stateless nations and dependencies – are often viewed in temporal terms: as vestiges of empire, as seemingly transitional and fleeting forms of politics, and also as increasingly durable: as polities 'stuck' midway on a path between recognised statehood and state collapse (Bahcheli et al. 2004; Clapham 1998; Spears 2004). Meanwhile, at a smaller scale, spaces of statelessness in the form of refugee camps are frequently referred to as sites of liminality and enduring temporariness, where time, but not history, is suspended (Agier 2002; Bauman 2002: 345; see also Chapter 4). This empirical study of how a polity like TGiE carves out an existence whilst 'out-of-place' in exile seeks to trouble these assumptions of the timelessness of statelessness and the teleology of state-building. To do so, I focus on how, through rehearsing a series of state practices, this polity holds in tension the protracted limbo of exile with the anticipation of possible futures.

The indeterminate state of being in exile should, on paper, be a temporary phenomenon. After all, the purpose is to return home, not to become established in a host state. However, in the case of protracted exile such as that of the Tibetans, this 'frozen' state can last for generations. This poses uncomfortable questions about the purpose of exile: questions posed eloquently by 'Gen Sherab' (a pseudonym, which highlights the sensitivity of airing such views) on the Tibetan news portal and web forum Phayul.com:

> the very goal of going into exile is but to refill our water bags and restock our armory so that we might return home to finish the fight with renewed vigor. But in the case of our exile, something else happened: we came, we saw, we stayed ... After fifty years of waiting, we are still here – essentially waiting. Waiting for what? (18 January 2007)[4]

Experiences of prolonged waiting are certainly not unique to political exiles. As Craig Jeffrey (2008) notes, the phenomenon of 'chronic waiting' is an increasingly common experience for a number of subaltern communities including asylum seekers (Mountz 2011), urban slum dwellers (Appadurai 2001) and the unemployed (Jeffrey et al. 2008). Paralleling some of the emotional connotations attached to exile, 'waiting' is conventionally associated with a sense of detachment, melancholy and frustration: of being trapped in a state of infinitely lingering and anxiety of being left behind. Yet waiting can also have a productive aspect. Illustrative of this is Finn Stepputat's (1992) work on Guatemalan refugees in which he traces how shared experiences of waiting generated 'a powerful sense of moral and political community that acted as a kernel for projects of opposition and social uplift' (Jeffrey 2008: 957). As I trace out in the chapters that follow, in the case of the TGiE waiting also has such a productive element, offering time and space for experimentation, preparation and reflection. Crucially this purposive waiting and the consequent framing of exile as a social and political resource is contingent on hope of potential futures different from the present.

Offering revealing lenses on this distinctive relationship between state practices and imagined futures are two strands of emergent geographical scholarship. The first is work on anticipatory action that interrogates ideas of the event, in particular looking at how the deployment of the term 'emergency' in relation to terrorism, climate change and epidemics opens up some form of exception to normal social and political life in liberal-democratic societies (Adey & Anderson 2012; Anderson 2010; Collier 2008; Duffield 2010). This body of scholarship explores the different anticipatory logics through which future emergencies are intervened on (such as pre-emption or preparedness) and the techniques and practices through which futures are known in order to be rendered actionable. Second, and approaching the idea of planning for possible futures from a different political perspective, is work on the notion of prefigurative politics. With roots in the New Left movements of the 1960s, prefigurative politics are modes of social action through which alternative – and often utopic – futures are enacted in the present, with the rationale that this makes that future more likely (Cornell 2011; Pickerill & Chatterton 2006). In its more contemporary incarnations, for example in the 'Occupy' movement (Sparke 2013; Vasudevan 2015), attention has shifted to 'being' in the present: to the politics of utopia for the moment, not necessarily with a particular future in mind. What both notions of anticipatory logics and prefigurative politics bring to the fore is how desired futures are imagined and practised in the present. In thinking through these modes of temporality via the case of exile Tibetan politics, this book therefore asks how imagined futures in the present play out in sites of displacement, in liminal spaces and in the margins of geopolitics.

The argument that focusing on the marginal offers an insightful perspective on the centre is one that has gained traction in anthropology and geography. Tim Cresswell makes a convincing argument for paying attention to 'the geographically marginal' in order to 'question the naturalness and absoluteness of assumed geographies. Like Bakhtin's carnivalesque-comic view of the world, these marginal events foster "'a realisation that established authority and truth are relative" (Bakhtin 1968: 10)' (Cresswell 1996: 149). Meanwhile in anthropology Yael

Navaro-Yashin (2003) uses the case of the 'no man's land' of the Turkish Republic of Northern Cyprus to argue for the exception having something to tell us about the rule, the abnormal about the normal.[5] Placing the marginal centre-stage is also at the core of Willem van Schendel's (2002) and James Scott's (2009) work on the idea of 'Zomia', the remote massif of Southeast Asia that was historically beyond state control. Exploring a more expansive notion of Zomia as a mode of behaving that rejects the state, Scott seeks to write an anarchist history from the margins: 'a history of deliberate and reactive statelessness' (2009: 3) that runs counter to the dominant narratives of state-making that are told from the political and cultural 'centres'.

Yet despite the revealing insights that focusing on the margins offers, such scholarship often finds itself being marginalised. In setting out an agenda for a postcolonial economic geography, Jane Pollard and colleagues express frustration at having to justify their focus on what are seen as 'quirky case studies... beyond the spaces of mainstream economic geography' (Pollard et al. 2009: 140). Meanwhile, confirming Jenny Robinson's observation that only knowledge generated in the 'heartlands' is seen 'as generative of theoretical and general geographical knowledge' (2003: 278), mainstream scholarship in international relations (IR) and political geography has largely overlooked the significance of non-state polities. Despite such polities seeking to operate within the international system, they are conventionally perceived as residing outside the territorial logic of sovereignty and placed alongside statist readings of the transnational stories that 'haunt the embassy, the law of the sea, the UN, the internet, international financial systems, and offshore economic zones. Such are exceptions that define the rule whose primary "author", "creator" and "guarantor" remains the state' (Soguk & Whitehall 1999: 679).

Central to such understandings of marginal polities as exceptions to the 'natural laws' of international society, is the assumption that there exists either absolute sovereignty or no sovereignty at all. Seen in relation to the 'norm' – the territorially bounded sovereign nation-state – such polities are consistently framed in negative terms, as *non*-states, *non*-sovereign, *un*recognised. In suggesting that it is more productive to focus not on what the TGiE lacks but on how it works, it is necessary to turn to more critical and expansive notions of the relationship between the state, sovereign authority and territory. In turn, engaging with this literature through the lens of polities like the TGiE brings to the fore questions around the relationship between sovereignty and territory, constructions of political legitimacy, the aspirational nature of statehood and the performance of statecraft.

'Unbundling' Sovereignty, Territory and the State

The bundling together of political authority, territory and population into the sovereign state remains the principal mode of conceptualising the organisation of political space both within a range of broadly realist scholarship (particularly in IR) and, thanks to the institutionalised dominance of this Western idea of the state, also within global popular imaginations. This view of the state as a 'container' for politics (Taylor 1994) is fundamental to both the legal criteria for statehood (e.g. The Montevideo Convention on Rights and Duties of States 1933, Article 1) and to dominant definitions of the state. These include Weber's sociological understanding of the state as 'a human community that (successfully) claims the *monopoly of the legitimate use of physical force* within a given territory' (1919/2009: 78, emphasis in the original), through Mann's (1984) designation as a set of centralised institutions that exercise power over a specific territory, to Giddens' definition as a 'political organisation whose rule is territorially ordered and which is able to mobilise the means of violence to sustain that rule' (1985: 20).

However, with state functions – from welfare to defence, economic regulation to law-making – increasingly both exceeding the territorial boundaries of the state and being articulated by non-state actors, there has been a drive to 'render mobile, fragile and contestable what traditional political discourses tend to naturalise' (Shapiro 1996: xxii). Instrumental to such a shift within the discipline of geography has been the field of critical geopolitics, with its aim of investigating and challenging the geographical assumptions that underpin the making of world politics. Whilst careful not to eschew the 'substantial material power of state institutions' (Dodds et al. 2013: 8) scholars of critical geopolitics nevertheless make a convincing case that political authority is not necessarily only state power, nor necessarily territorially constituted (Agnew 2005).

One of the most influential critiques of the coalescing of bounded territory and political power in the form of the sovereign state is John Agnew's (1994) framing of it as a three-pronged 'territorial trap': an ahistorical reification of states as fixed units of sovereign space; a view of the state as the pre-existing container of 'society'; and a dichotomising of the domestic (inside) and the foreign (outside). In a similar vein, a number of scholars have argued that these traditional constituent elements of sovereign statehood so often conflated in political analysis are subject to 'unbundling'. Following John Ruggie (1993), James Anderson argues that we are seeing the 'dispersal of authority to different types of institutions at different levels' (1996: 140) reminiscent of the vertically segmented and overlapping authority that existed in Europe before the rise of the modern state. Similarly, in their volume *State Sovereignty as Social Construct*, Biersteker and Weber (1996: 19) assert that disentangling the concepts of state and sovereignty offers the 'theoretical possibility of non-sovereign territorial states (Taiwan) or the existence of sovereign non-territorial ones (Palestine)' thus enabling us 'to consider whether entities other than territorial states can begin to make legitimate claims of final authority'. As such, this expanded view of how power and space are organised opens up valuable conceptual spaces in which the functioning of geopolitical 'anomalies' such as the TGiE can come into view. However, whilst using this unbundling of state and sovereignty as a starting point for examining the everyday functioning and political aspirations of the TGiE, this book goes further and also examines the deliberate reassembling of statehood and political legitimacy in contexts of liminal territoriality and legality.

Turning to the concept of sovereignty, critical rethinking of this cornerstone of political theory also provides important theoretical leverage for examining how polities like the TGiE function and what their existence means for understandings of politics at a range of scales. Within the post-structuralist inspired critical turn in political geography, political anthropology and the fringes of IR, sovereignty has been conceptualised as historically contingent (Barkin & Cronin 1994; Krasner 1999), socially constructed (Ashley 1988; Walker 1993) and negotiated through performances, discourses and everyday materialities (Edkins et al. 1999; Hansen & Stepputat 2005; C. Weber 1995). Not only does such scholarship conceive sovereignty as divisible and incremental (Agnew 2005), but it also enables key questions to be asked: what cultural and political resources are mobilised in order to make claims of sovereignty (Duara 2003)? If polities 'can lose sovereignty, can they also gain it?' (Elden 2006: 18). What is it that is sovereign: an abstract notion of a state or polity, or the practices of its officials and 'citizens'? And can two sovereignties overlap and coexist within the same territory? It is through the empirical lens of polities aspiring to statehood that such questions gain both empirical and political purchase.

In posing such questions in my research on the TGiE I have argued that this polity articulates a mode of de facto sovereign authority (McConnell 2009a). Distinct from de jure sovereignty, which is based on recognition by other sovereign states and legal coercive powers, de facto sovereignty is an oft ignored route to sovereignty based on claims to and constructions of legitimacy. Building on John Agnew's (2005) work, I suggest that if we analyse how de facto sovereignty is articulated on the ground then not only does this refute the idea of sovereignty as absolute, but it also opens up space to examine what polities like this exile government actually *do*. Central to this is giving due consideration to the notion of political legitimacy. As Alice Wilson and I argue in a comparison of the TGiE and SADR (Wilson & McConnell, forthcoming), disentangling claims to and constructions of legitimacy from full legality as a recognised state brings to the fore the ambiguous nature of this concept: an ambiguity that makes it 'good to think with' (cf. Shain 1989). As I explore in the chapters that follow, liminality – both territorially in terms of displacement and legally in terms of lack of full recognition – can counterintuitively provide creative grounds for producing legitimacy and, seen through the lens of polities like TGiE, legitimacy thus emerges as not so much an achieved status, but as a set of techniques of government.

The drawing of a distinction between de facto and de jure authority also underpins the post-1945 shift in international norms that determine the nature and legal recognition of statehood. In essence this entailed a transition from recognition being a legal act that was sanctioned only after successfully demonstrating capacity to govern – the so-called doctrine of effective control (Murphy 1999) – to a juridical approach to statehood whereby 'rulers can acquire independence solely by virtue of being successors of colonial governments' (Jackson 1990: 34). This led to the establishment of quasi states that are 'internationally recognised as full juridical equals ... yet which manifestly lack all but the most rudimentary empirical capabilities' (Pegg 1998: 3) and to their flipside: de facto states that have state capabilities but lack recognition. As contenders to govern territory currently under the control of a (presumably) recognised state, governments-in-exile further expose both the zero sum game of recognition vis-à-vis territorial sovereignty, as well as the arbitrary and political nature of this practice of international law. We therefore see recognition frequently being granted, denied, withdrawn and reinstated to deposed and aspirant governments as geopolitical situations and relations evolve (Talmon 1998).

Returning to the elements that are so frequently bundled together in understandings of the state, it is territory – and the relationship between territory and political authority – that marks out exiled administrations like the TGiE as being particularly unusual and revealing cases. Perhaps the most striking feature of the TGiE is the fact that, whilst denied jurisdiction over territory both in the homeland of Tibet and in the host state, the TGiE's state-like functioning in India is seemingly contingent on its de facto control over the series of spatially dispersed Tibetan settlements (see Chapter 4). As such, this case questions the implicit assumption that 'the spatial extent of state sovereignty is coterminous with territory' (Painter 2010: 1095). Conventionally conceived as bounded and contiguous, territory is presented as an essential element in classic definitions of the state and also underpins the modern doctrine of state sovereignty in international law, with sovereignty taken to be a 'political legal fact within an already given and demarcated territory, simultaneously signifying sovereignty over the same territory' (Bartelson 1995: 29).

Given the above discussion it should come as no surprise that, within critical traditions, such absolute understandings of territory and its predetermined relation to state power are disputed both empirically and conceptually (Agnew 1994). We do not need to look far for contemporary or historical examples of shared sovereignty, overlapping jurisdictions, and cases where the principles of self-determination and territorial integrity are contested and violated. In line with the broader spatial turn within the social sciences there has been a proliferation of scholarship reflecting on the increasing spatial complexities of contemporary politics. This has included work on the contingency of the spatialities of state formation (Brenner et al. 2003; Painter 2006) and a sensitivity to the 'the plasticity of state spatiality' (Painter 2010: 1095) as well as understandings of territory as a process as opposed to merely a location (Forsberg 1996). It is precisely within this tension between territory as fixed/fluid and exclusive/overlapping, that the TGiE operates. As I explore in detail in Chapter 4, the TGiE's state-like territorialising strategies and innovative use of symbolic territory offer a revealing spotlight onto the continual production of territory through social relations, governance practices and imaginaries. I therefore use this case to raise questions about the persistent connection between territory and governance and to think creatively about the contingency of (non)state formation and its implications for the host state at a range of scales.

State(less)ness

Having thus far focused on the productive unbundling of connections between state, sovereignty and territory, I now want to focus in on the state itself and return to the two questions that underpin this book: *how* does the TGiE enact state-like functions in India, and *why* does it put so much work into emulating this form of political organisation? In asking whether the TGiE is a state-like polity or not, the perhaps flippant answer is that it depends on the theories of the state that are bought into. If we are to follow the classic institutional approach to the state based around Max Weber's assertion that a state must meet a defined list of criteria including a stable government with a bounded territory, a self-defined people and a monopoly of the legitimate use of physical force, then this polity will simply never count as a state. Likewise, given the fact that no state recognises the TGiE, it conclusively fails to meet legal definitions of the state as a juridical entity of the international system and international law (Montevideo Convention 1933, Article 1).

Legal readings of the state pose another obstacle vis-à-vis this case: the confla-tion of state and government. In a robust critique of political geographers' inter-changeable use of 'state' and 'government' (e.g. Agnew 1998; White 2004), Heath Robinson (2013) argues that this has hindered the development of state theory within the discipline, and calls for a more sustained engagement with interna-tional legal scholarship. And, indeed, from such a perspective the 'legal position of governments-in-exile is dependent on the distinction between government and state' (Crawford 2006: 34). However, the distinction is somewhat more uncertain in the Tibetan case than, for example, wartime governments-in-exile (Talmon 1998). On the one hand, with the TGiE being the re-establishment of the pre-1959 government of Tibet rather than a new entity established in exile, this clear continuity means that it exercises the legal personality on behalf of the former state of Tibet (Tibet Justice Center 2011). Yet on the other hand, and discussed in more detail in the following chapter, the prior legal recognition of the state of Tibet is disputed and it certainly no longer exists in any meaningful sense. Com-pounding this elusiveness is the fact that the TGiE has no formal legal status, and is no longer calling for international recognition. So, in strict legal terms, what we have in the TGiE is neither a state *nor* a government. However, as I will explore in detail in the chapters that follow, the TGiE nevertheless enacts a series of practices of government that bolster its claim to legitimacy and go some way to constituting a state-*like* polity in exile. If 'governments are not states, but they can represent them' (Robinson 2013: 560), then perhaps what we are seeing here is the TGiE representing a future possible state in the present.

In essence, pursuing legal definitions of state and government significantly limits what we can say about the TGiE, and offers little by way of addressing the questions of how this polity functions and why it emulates state-like practices. A more productive line of enquiry is opened up if we turn to the now sizeable body of critical sociological, geographical and anthropological approaches to the state. Engagements with social theory and detailed empirical research on contemporary articulations of statehood have, in many ways, revitalised political geography and political anthropology, 'producing innovative and vital accounts of state operations, imaginaries and effects' (Feldman 2008: 233; see also Hansen & Stepputat 2001, 2005; Jeffrey 2013; Jones 2007). Central to such approaches is an acknowledgement that classic institutional definitions of the state, like Weber's, are rarely realised in practice, and an understanding of the state as an 'elusive object of study, difficult to approach both practically and conceptually' (Bernstein & Mertz 2011: 6). Such scholars question the supposed unity of the state and, 'in place of coherence and consistency of purpose, they find state activities to be chaotic and incoherent assemblages of sites, processes and institutions' (Nugent 2007: 198).

However, the project of questioning the existence of the state as a distinct social actor is neither as recent nor as wedded to post-structuralist perspectives as is perhaps first imagined. We can, for example, trace it to the 'legitimacy approach' to state-building, which, influenced by Durkheimian sociology, emphasises that the state comprises the collective beliefs and ideas that a society has fostered over time (Lemay-Hébert 2009). Drawing on such approaches IR scholar Barry Buzan emphasises the socio-political rather than physical dimensions of the state: 'in some important senses, the state is more an idea held in common by a group of people, than it is a physical organism' (1991: 63). Meanwhile, influential within sociology and political geography is Nicos Poulantzas' concept of institutional materiality whereby the state should be regarded not as an intrinsic entity but as an institutional ensemble:

a relationship of forces, or more precisely the material condensation of such a relationship among classes and class fractions, such as this is expressed within the State in a necessarily specific form... by grasping the State as the condensation of a *relationship*, we avoid the impasse of that eternal counter position of the State as a Thing-instrument and the State as Subject. (Poulantzas 1978: 128–129; emphasis in the original)

Extending Poulantzas' argument of the state as a form of social relation, Bob Jessop's (1990) strategic relational approach defines the state as a 'specific institutional ensemble' without a pre-given unity or fixed boundaries. As Painter and Jeffrey note, not only does this approach promote the idea of states as multifarious in their activities but it also undermines the assumption that there exists an essential distinction between state and non-state institutions (Painter & Jeffrey 2009: 26). A similar argument is made by anthropologist David Sneath in his fascinating account of state formation in the steppe societies of Inner Asia. He writes that:

> Viewing the state as a form of social relation rather than a central structure avoids the evolutionist dichotomy between state and nonstate society, and it makes it possible for us to [...] look for 'state relations' in societies that do not seem to match the older models of the centralized, clearly bounded state. (Sneath 2007: 1–2, 10)

Central to these relational approaches to the state is a rethinking of the nature of power that draws on Foucault's scholarship. Essentially this is a shift to seeing power not as absolute, centred and something that can be possessed and thus lost, but as diffused and networked: a relational effect rather than an attribute or property (Allen 2003; Foucault 1977, 1991).

Foucault's engagement with the concept of the state is complex. In his earlier work, he was renowned for his criticisms of state theory, rejecting outright a state-centred notion of power and advocating instead a bottom-up approach (e.g. *Discipline and Punish*: Foucault 1977). This was developed further in his writings on governmentality and biopolitics. As a broader category than 'government', governmentality – the organised practices through which we are governed and through which we govern ourselves – 'breaks with many of the characteristic assumptions of theories of the state, such as problems of legitimacy, the notion of ideology and the questions of the possession and source of power' (Dean 1999: 9). In thus eschewing the assumed co-dependence of territory and sovereignty and displacing the notion of a unitary state, Foucault's conceptualisation of governmentality opens up the possibility of diverse and heterogeneous agencies controlling the mechanisms of authority (Dean & Henman 2004: 483; see also Ong 1999; Rose-Redwood 2006). However, Foucault's later lectures, *Sécurité, Territoire, Population* (Foucault 2007), mark 'a decisive turn ... to interest in changing forms of statehood and statecraft' (Jessop 2007: 34). Within this work Foucault presents the state as 'an emergent and changeable effect of incessant transactions, multiple governmentalities, and perpetual statizations' (Jessop 2007: 36–7; see also Foucault 2004: 79). Or, simply put, the 'state is not the source of power but its effect' (Marston 2004: 4). As I discuss in Chapter 6, the notion of governmentality, and Foucault's articulation of the relationship between political authority, territory and population, proves to be an instructive framework for exploring how a territory-*less* polity such as TGiE enacts governance practices. In turn, this exile government's transgression of the distinction between state and non-state, legality and legitimacy, opens up space to consider the notion of governmentalisation to practise the state (see also McConnell 2012).

Writing at around the same time as Foucault, but across the English Channel, sociologist Philip Abrams proposed an arguably more radical approach to the state by challenging its very ontology. In an address to the British Sociological Association in 1977 (published in 1988), Abrams laid out a powerful critique of conventional state theory for reifying the state and instead argued that conceptualising the state as an *idea* rather than a political or material reality draws attention to the continual processes through which it is reproduced:

> we should recognize [the] cogency of the idea of the state as an ideological power and treat that as a compelling object of analysis. But the very reasons that require us to do that also require us not to believe in the idea of the state, not to concede, even as an abstract formal-object, the existence of the state. (Abrams 1988: 79)

Following this argument, the 'difficulty of studying the state' (Abrams 1988: 79) lies in the fact that as a unified political entity it simply does not exist. A key theoretical legacy of Abrams' argument is the questioning of the subjectivity of the state. Rather than a pregiven subject of international politics, the state under such readings is understood as being produced through dynamic social processes and thus is constantly in emergence. Given the nascency and limbo status of the TGiE as a government 'transplanted' to exile, this case offers a particularly instructive insight into the contingency of statehood.

Related to a focus on the processual nature of the state has been a turn towards the role of everyday state practices. Extending Abrams' argument and pushing further Foucault's notion of the state as a structural effect, Timothy Mitchell argues that the state 'should be examined not as an actual structure, but as the powerful, metaphysical effect of practices that make such structures appear to exist' (1991: 94). Thus, whilst for Abrams the state was solely an 'ideological construct' – 'what exists is the belief that the state exists' (Bratsis 2006: 13) – Mitchell argues that the state does have a *social* existence and therefore attention needs to be paid to the materiality of the state as a structural effect. In conceiving the state not as something concrete there to be observed but rather as a structural effect constituted of everyday practices, then not only does the TGiE emerge as a viable focus of enquiry, but we are able to start to get a handle on *how* this polity functions in exile.

With political anthropologists – and increasing numbers of political geographers – seeking empirically to investigate everyday practices through which state ideas are reproduced, two conceptual framings are of particular interest here. First, and explored in more detail later in this chapter, the micropolitics of everyday state-making has seen a productive engagement with Judith Butler's (1990) notion of performativity. Seen through the lens of performativity, the state is a construct evolving through a continuous performance of norms and practices that constitute the effects of the material state (Constantinou 1998; C. Weber 1995,

1998). Second, and relatedly, is a growing interest in the ethnographic study of the state (Ferguson & Gupta 2002; Herzfeld 1992; Jeffrey 2013; Trouillot 2001). By attending to 'the local, the emic, the vernacular notions of governance' (Hansen & Stepputat 2001: 9) such work advocates a more disaggregated study of the state: one that highlights the production of arbitrariness (Gupta 2012) and where sites of everyday life become 'a central domain for the production and reproduction of the state' (Navaro-Yashin 2002: 135). This turn to the banal and prosaic processes through which states are reproduced (Billig 1995; Painter 2006) means that 'sites less obvious than those of institutionalized politics and established bureaucracies' (Trouillot 2003: 95) become valid perspectives from where the state can be studied in terms of its effects. In the context of an exiled administration operating out of a host state, it is precisely such unconventional sites where the everyday techniques of government – naming, reporting, taxing, authorising – are enacted, which is of interest here.

Grounded in this work on the everyday state, Joe Painter's exploration of the notion of stateness is a framing that guides this study in a number of ways. Posited in distinction to the more restrictive concept of state*hood*, which 'refers almost exclusively to the recognition of a state in international law' (Painter 2006: 755), state*ness* is the various social relations and mundane practices that give rise to state effects. As Painter notes:

> the actualization of stateness in particular sites through particular kinds of practices and encounters clearly involves considerable work... This effectuation of the state requires energy, time, knowledge and expertise on the part of state officials, but also (as Corbridge et al. 2005 reveal) on the part of citizens. (Painter 2007: 606)

In bringing the notion of stateness into dialogue with the temporalities of exile – of waiting and anticipating futures – this study therefore offers a fresh angle on the state: one that sharpens into focus the different temporalities through which state effects are generated, the tension between the state existing and emerging and the rehearsed nature of state practices. In addition this intersection also provides a provocation for the question of *why* an exiled community, scattered across multiple states and with limited human and financial resources, should seek to invest in such labour to produce a state-like polity as a set of effects.

In seeking to address this question, we need to look to the discursive and symbolic aspects of the state as well as its actualisation through everyday practices. As Akhil Gupta has detailed in his analysis of state corruption in India, focusing on the discursive construction of the state draws attention 'to the powerful cultural practices by which the state is symbolically represented to its employees and to citizens of the nation' (1995: 377). In their edited volume on ethnographies of the postcolonial state, Hansen and Stepputat note that the 'language of stateness' (2001: 9) – the production of narratives of state power and the construction of myths of the state – is an increasingly important mechanism through which the state emerges as an imagined collective actor. On one level this calls for a focus on the politics of inscription (Gupta 2012). In light of the material and legal limitations that aspirant states and governments-in-exile face, examining what the 'state states' (Corrigan & Sayer 1985) is arguably more important for such polities than it is for established nation-states. As I discuss in Chapters 5 and 6, the texts of governance, and the knowledge-generating techniques that underpin them, are a revealing lens onto how the TGiE defines and presents itself, its relationships with the host state and its place in the world. However, there is more than simply bureaucratic representations behind the powerful myth of the state. Key to understanding the state – and cases of non-state stateness – are questions of idealisation and affect.

The idealisation of the state as a mode of politics to aspire to and to mimic runs somewhat counter to the relations of apathy, distrust and disillusionment that many living in 'prized' Western liberal democracies seem to articulate. Offering one explanation for the continued romanticisation of the state in the midst of such negative narratives is a growing body of work on affect. While the 'subjective dynamic that sustains the state as a powerful, inescapable, social reality' (Aretxaga 2003: 400) has been a focus of attention for some time – Weber himself understood that obedience to the state is determined by powerful motives of fear and hope (Weber 1919/2009) – recent work in anthropology has highlighted the specific role that emotions and passions play in the everyday state (Navaro-Yashin 2012; Stoler 2007; Taussig 1992). The contribution of this book to such literature on the state and affect is to add empirical depth to the role of the state as a screen for political hopes, desires and aspirations for empowerment and to think critically about intersections between the politics of hope and the distinctive temporalities of exile. As I explore in the chapters that follow, hope in the exile Tibetan context cross-cuts both the question of return at the scale of the nation, and everyday practices of carving out a Tibetan way of life in exile. As such, it provides a conceptual bridge between notions of stateness, the state ideal and state affect on the one hand, and the distinct temporalities of exile on the other: 'The taking place of hope enacts the future as open to difference but also reminds us that the here and now is "uncentered, dispersed, plural and partial"' (Gibson-Graham 1996: 259; see also Anderson 2006; Miyazaki 2004).

At first glance discussions of hope and visions of utopia in the context of a situation of forced displacement and prolonged exile seem inappropriate at best and callously trivialising at worst. Edward Said, who was famously outspoken on the dangers of romanticising exile, rightly warned that 'to think of exile as beneficial, as a spur to humanism or to creativity, is to belittle its mutilations. Modern exile... has torn millions of people from the nourishment of tradition, family and geography' (Said 1984: 50). And yet to focus solely on the alienation, dispossession and despondency of exile (Goddeeris 2007) strips exilic communities, individuals and institutions of agency and aspiration. This is particularly the case for exiled communities who are striving to raise the profile of their case on the international stage in order to garner political, moral and material support. Recounting the atrocities being enacted in the homeland and the challenges of life in exile is, of course, important, but international audiences have grown blasé to tales of woe and fervent nationalist causes. As I discuss in Chapter 7, proving your worth as global citizens in the present and demonstrating aspirations to practise democratic good governance in the future are also crucial legitimising strategies for exile struggles.

Performing and Rehearsing the State

> Learning through doing, repeating for mastery, and improvising with given circumstance – the hallmarks of rehearsal – were intended to lead to a performance (someday, if necessary, but for the time being deferred) in the sense that performance is a naturalised execution of an uninterrupted unfolding sequence of actions. This does not require acting skill, but it is acting; this is not theatre, but it is theatrical; this is not performance but it has a methexic relationship to what could someday be performed.
> Davis 2007: 88

As Alex Jeffrey notes, 'the theatre has provided the social sciences with a rich set of terms and metaphors through which to narrate the complexity of social and political life' (2013: 26). Amongst terminology such as 'audience', 'actor', 'stage' and 'mimicry' it is the concept of performance that has been most pervasive and influential: 'according to many scholars, everyday acts are performance, and indeed nothing or very little may escape the rubric of the performative' (Davis 2007: 84). Given the diversity of scholars engaging with the notion of performance, the analytical insights it is claimed to provide vary considerably according to the different conceptions of power, subjectivity and agency with which it is brought into dialogue. An early and influential school of thought on performance emerged from work on symbolic interactionism in sociology. Directly appropriating and adapting vocabulary from the theatre, sociological models of dramaturgy analysed everyday social interactions as if they were theatrical performances, with social life thus being conceived as 'staged by conscious agents who adhere to scripts' (Pratt 2009: 525). A pioneer in this field was Erving Goffman, whose text *The Presentation of Self in Everyday Life* uses the language of the stage – 'of performers and audience, of routines and parts; of performances coming off or falling flat; of cues, stage settings and backstage' (Goffman 1959: 246) – to explore the strategies through which individuals manage themselves in the presence of others. A rich analysis is thus provided of the relationship between performance and audience, reality and contrivance, as well as the distinction between 'front regions' where performances are enacted and observed, and 'back regions' where an individual's appearance, manner and performance is prepared.

Whilst Goffman's ideas have been productively applied and reworked by geographers through grounded studies of public and private performances enacted in specific work places (e.g. Crang 1994; McDowell & Court 1994; see also Davidson 2003), his framework has been the subject of significant critique. This has focused primarily on the nature of subjectivity assumed by the dramaturgical analogy: an assumption that, pre-existing any performance, there are conscious, motivated agents. Compounding this problematic separation of the performer and the performance is the implication that the researcher assumes a (masculinist) directorial role 'with a privileged knowledge of the scripts, costumes and comportment of the "actors" involved' (Jeffrey 2013: 28). Cognisant of these critiques, yet retaining a focus on intersections between performance, identity and social interaction, is a growing body of work that attends to the notion of performativity. Acknowledging that 'performance is subsumed within and must always be connected to performativity' (Gregson & Rose 2000: 433), engagement with the latter has, in broad terms, meant a shift from a theatrical to a linguistic approach and, crucially, a questioning of the idea of a stable subject. The work of gender theorist Judith Butler (1990) has provided inspiration and a point of departure for many scholars pursuing this line of enquiry. Distinguishing performativity – 'the reiterative and citational practice by which discourse produces the effects that it names' (Butler 1992: 2) – from a singular or deliberate 'act' of performance, Butler draws on and extends a Foucauldian understanding of the relationship between power and knowledge. The idea of iterative practices bringing into existence and naturalising particular social categories and modes of being has been influential within and beyond feminist and queer theory in rethinking the construction of power relations, identity and social difference. For example, geographers Nicky Gregson and Gillian Rose make a persuasive case for thinking about space as performative: 'performances do not take place in already existing locations ... specific performances bring these spaces into being' (Gregson & Rose 2000: 441).

Returning to the core concerns of this study, what does the idea of performance add to understandings of politics and the political? Performative accounts of subjectivity in relation to sovereignty and state action – particularly in the form of foreign policy – have proved to be a productive line of enquiry in the critical strand of IR. Contra realist and neo-realist assertions, states are thereby understood not as pre-given subjects, but as evolving processes of repeated actions, behaviours and practices that collectively create the image of a stable national identity (Campbell 1992; George 1994). As Cynthia Weber argues, such analysis of states as performative 'moves beyond traditional definitions of sovereignty ... To what must a state "do" in order to "be" sovereign?' (1998: 92).

The lens of performativity is thus productive in terms of destabilising categories and troubling the notion of political subjectivity. However, the comparative sidelining of the *theatrical* notion of performance in this scholarship loses sight of two elements that are key to this study. First is the theatricality of formal politics: the public show of 'key metonymic images of nationalism and nationhood' (Hansen 2001: 226) and the symbol-laden nature of 'imagineered' ceremonies (Ley 2000) such as national days and meetings of state leaders. James Sidaway's account of the constitution of regional communities through the ritual of diplomacy on the international 'stage' is instructive here. As he notes, 'the exploration of... statecraft as drama or theatre impressed itself upon me through my attendance at formal international events... amidst introductions, appeals, applause and speeches' (Sidaway 2002: xi). Second is the deliberate nature of the exile Tibetan political project: this is a purposeful performance of stateness to seek legitimacy. The work of cultural anthropologist Victor Turner, a figure key in the development of a dramaturgical perspective but largely overlooked in political geography and IR, can offer insights in this regard. Where Goffman takes a scenographic approach to human interactions – all social encounters are staged – Turner (1974, 1987a) focuses attention on the performative nature of specific 'social dramas' that emerge from situations of conflict. In his words, 'the dramaturgical phase begins when *crises* arise in the daily flow of social interaction. Thus if daily living is a kind of theatre, social drama is a kind of metatheatre' (Turner 1987a: 76). Turner's close collaboration with drama theorist and avant garde theatre producer Richard Schechner spurred particularly innovative analyses of the mutually reinforcing interconnections between ritual and theatre, social drama and aesthetic performances. Though critiqued for his nostalgia for the more sacred life, Turner's approach to performance as 'the art that is open, unfinished, decentred, liminal... a paradigm of process' (Schechner 1987: 8) is a productive one, and his analysis of the creativity that can arise in situations of liminality – when conventional structures are suspended – has particular resonances with the case of governance in exile.

The approach adopted here revisits theatrical notions of performance and state ritual and brings them into conversation with the critical insights that performativity lends to politics and the political. Social theorist Jeffrey Alexander makes a persuasive call for such an approach in his examination of socio-political performance as existing between ritual and strategy:

> If we are to understand how power is exercised, we need not only cultural theory from... hermeneutics, semiotics and post-structuralism, but the tools of theatre, film, television and performance studies, of media research, of reception theory and ideas about emotion and materiality. (Alexander 2011: 4)

Another productive example of the bringing together of theatrical notions of performance with Butler's understanding of performativity is Paul Higate and Marsha Henry's account of the spatial practices and security performances of peacekeeping operations in Haiti, Kosovo and Liberia. In tracing how peacekeeping is staged and perceived through 'its enactment as an embodied performance that necessarily constituted an "audience" which required securing', Higate and Henry demonstrate how the concept of performance 'sensitises' us to the everyday and banal aspects of daily life in these peacekeeping situations (2009: 17).

This study takes inspiration from such scholarship, which navigates between performance and performativity, but it is the notion of rehearsal rather than performance that has more precise resonances with the political project that is the TGiE. The distinction between performance and rehearsal is a slippery one. Theatre studies has conventionally perceived a threshold between the two, with the presence of an audience on opening night rendering performance an experience on a different register to that of rehearsal. Indeed, as Turner notes, the 'etymology of "performance" ... has nothing to do with "form", but derives from Old French *parfournir* "to complete" ... A performance, then, is the proper finale of an experience' (Turner 1982: 13). In contrast, performance studies sees slippage between them: theatre *is* performance but rehearsal may be seen *as* performance (Schechner 2002: 41–42). In seeing rehearsal as *performative*, my approach follows Tracy Davis's argument that 'rehearsals have different framing devices than performances' and thus 'rehearsal is a viable category for explaining an empirical testing-out' (2007: 254, 88).

Davis's exploration of rehearsal is particularly compelling. As a historian of theatre and performance theory she focuses on the 'conscious reference to the theatrical mode of rehearsal' (Davis 2007: 84) within Cold War nuclear civil-defence planning in the United States, Canada and the United Kingdom during the 1950s and 1960s. Davis thus examines the activities through which citizens and state leaders rehearsed for nuclear war – evacuating cities, treating the wounded (played by actors) and broadcasting emergency messages – and asks 'how did people take part, who was on the sidelines, and how did they anticipate an eventual – though perhaps perpetually deferred – performance?' (Davis 2007: 2). Though suggesting intriguing parallels in terms of anticipatory logics, the articulation of rehearsal in the case of civil defence diverges in important ways from the focus of investigation here. Aside from the obvious – the imagined future in the exile Tibetan case is a hopeful one of return, rather than a nuclear apocalypse! – the rehearsal explored in this example is not one that is continually articulated as such. Whilst the experimentation and practising of governance techniques is certainly discussed by TGiE's leadership and bureaucrats, the functioning of this exiled administration is not first and foremost a simulation or scenario exercise: it is not, as per civil defence, 'an embodied mimetic methodology that is *inherently* and *crucially* theatrical' (Davis 2007: 2; emphasis in the original).

Yet, the theatrical notion of 'rehearsal' nevertheless offers a series of framings that are crucial to what I am seeking to do in this book. First is to situate this study precisely in dialogue with the rich body of research outlined above which employs dramaturgical terminology to explore the processual and contingent nature of political identities, sovereign authority and state actions. Rehearsal, I suggest, both cross-cuts these strands of thinking on performance and performativity, and brings to the fore the provisional and pedagogical nature of TGiE's stateness. Second, rehearsal is used to provide a form of narrative through the chapters that follow. As the discussion in this chapter has indicated, this book seeks to bring a diverse range of concepts and approaches into conversation: a processual rather than categorical approach to polities that aspire to play the state game; the temporalities of exile in terms of prolonged waiting and anticipating futures; a critical re-examination of the relationship between sovereignty, territory and the state; and the state as constituted through everyday practices, relations and affects. Whilst imposing a neat conceptual framework on such ideas would run counter to the overarching approach endorsed here, some form of integration is nevertheless needed. The notion of rehearsal plays that role.

Finally, and most importantly, rehearsal acts as a provocation. Whilst it is a concept used with caution and moderation in this book – like any framing it has the potential to distract attention and obscure certain issues – rehearsal nevertheless enables novel questions to be asked of this case of state(less)ness. One such set of insights that the lens of rehearsal offers is into the significance of training, practice and professionalisation in the field of governance that has become so central to the TGiE's self-declared purpose. Just as *stage*craft must be learned, honed and practised, so must *state*craft. Rather than strategic relations between different states – conventionally through the instruments of diplomacy and war – statecraft has been reinterpreted within critical geopolitics as practices that make the state and its importance seem both natural and necessary (Campbell 1992; Doty 1996; Kuus 2008). Drawing on this work, I also push the term further and focus on the idea of crafting the state: as state-*craft* as a set of practices that have to be learned and perfected and an art that, in turn, (re)constitutes the idea of this 'state'-in-exile. Rehearsal is therefore a lens that offers an alternative spotlight on exile Tibetan politics – one that highlights the creative potential for prefiguring Tibetan statecraft – and opens up space for critical reflection on the temporal and spatial nature of state practices more generally.

Endnotes

1   For scholarship that attends to the statelessness of the Tibetan case see Anand (2003), Garratt (1997) and Hess (2006).

2   When discussing polities that do not fit the traditional form of the modern state, I am not referring to the likes of transnational companies, supranational organisations, extra-state criminal cartels or NGOs. Whilst these political actors often share characteristics of territorial dispersal and formalised bureaucracies, what distinguishes the TGiE is its aspiration to state-like functioning and its emulation of state practices.

3   Finding an appropriate label to describe the polities discussed here is a tricky task and there is a lack of consensus over terminology amongst scholars working on such cases. My own use of such terms is loose in nature, and I suggest that the metaphor of rehearsal offers a possible alternative that avoids problematic binaries and hierarchies.

4   Phayul.com website. LHASA CALLING: Why we must escape from exile (www. phayul.com/news/article.aspx?id=15363&article=Escape+from+Exile).

5   Whilst Navaro-Yashin invokes Giorgio Agamben's concept of the 'state of exception' (Agamben 2005), it is not a notion that I bring into play in this study. With the TGiE's absence of legal jurisdiction this is quite the opposite of sovereignty in its final instance and a limit case 'which throws into crisis the original fiction of sovereignty' (Agamben 1995: 118).

Chapter Three
Setting the Scene: Contested Narratives of Tibetan Statehood

The exiling of a government and its re-establishment in a foreign state is invari-
ably contingent on contested and often violent historical contexts and political
events. As such, any appreciation of the role and functioning of a contemporary
government-in-exile needs to be grounded in the circumstances under which it
was established and the political institutions that the exiles brought with them, as
well as the nature of the host state environment and the aspirations of the exiled
leadership. The purpose of this chapter is to set just such a scene for what is to
come in this exploration of the TGiE's rehearsal of stateness. The backstory will
be filled in, the plot and main protagonists introduced, and some pointers given
as to where the 'drama' might head in the subsequent chapters.

The chapter starts with two 'preludes' to the twentieth-century crisis in Tibet
which introduce the historical and political context of this region through the
lens of theoretical concerns outlined in the previous chapter: the nature of the
state and the relationship between territory and authority. The first section sets
out the characteristics of the Tibetan polity from the founding of the Tibetan
Empire in the seventh century, through to the increasingly consolidated nature
of Tibetan stateness at the end of the nineteenth century. In the second prelude
the focus turns to the relations that the Tibetan government fostered with its
neighbours during different periods, and emphasises two key points of contestation: delimiting the spatial extent that is 'Tibet', and defining the nature of the
power relationship between these political actors. Whilst pre-twentieth-century
Tibet is presented in broad brushstrokes, comparatively more fine-grained detail
is provided of recent Tibetan history. For it is the periods of de facto statehood

Rehearsing the State: The Political Practices of the Tibetan Government-in-Exile, First Edition. Fiona McConnell.
© 2016 John Wiley & Sons, Ltd. Published 2016 by John Wiley & Sons, Ltd.

(1911–1949), Chinese annexation (1949–1951) and the circumstances leading to the flight of the 14th Dalai Lama and his government into exile in 1959 that are not only fundamental to the existence of the TGiE but are also the periods that the exile community most frequently refer to in discussion on the present and future status of Tibet. The final part of the chapter brings the backstory up to date and sketches out the form and evolution of the TGiE, the key periods of Tibetan migration to South Asia, and the continuities and disjunctures between the exile government and its pre-1959 manifestation.

Prelude I. Pre-modern Tibet as a 'Stateless Society'?

The simple question 'What is Tibet?' proves to be a disconcertingly difficult one to answer. Descriptors are not only politically charged, but are also often inaccurate and misleading. Particularly problematic is attaching modern Western notions of the nation-state to various political structures in this region at different periods of history. The model of a centralised state with absolute authority enacted within precise borders and over a unified national population has been a largely unrealised ideal in the modern Western European context, and, as discussed in the previous chapter, it therefore has significant limitations for examining state forms more generally. It is also a model that, as Giddens (1985) notes, is a particularly poor fit for pre-modern states where authority was geographically uneven and overlapping, as allegiances of subordinate rulers shifted over time and space. In large parts of Asia it was well into the twentieth century before a centralised, territorial nation-state became the dominant political configuration. Before then, polities in the region often lacked a stable monopoly of power, and borders – or 'frontiers' as Owen Lattimore termed them – were shifting, porous and indeterminate spaces of interaction (Nordholt 1996). As set out in Lattimore's (1940/1949) extensive analysis of 'China and its marginal territories', this was a region dominated by cycles of dynastic and 'tribal' history until the rise of the Chinese state in the early twentieth century.

The unsuitability of the territorialised nation-state descriptor for Tibet did not, however, deter Western colonial officers and scholars from portraying the region as an exoticised 'theocratic variant of the centralized autocratic states familiar from elsewhere in Asia' (Samuel 1982: 215), with the Dalai Lama as a 'Living Buddha' possessing absolute religious and political authority over the Tibetan population and territory (e.g. Bell 1924; Chapman 1938). These 'Lhasa-centric' accounts from the early twentieth century assumed that this isolated and rigidly stratified Lamaist state had been the enduring political structure in this region and thus proceeded to project this model back into history. Similar trends of framing pre-modern political orders in terms of nation-statist and evolutionist logics can also be traced in Western scholarship on Inner Asia (e.g. Barfield 1989; Krader 1963). Recent historical investigations in both contexts have, however, convincingly demonstrated that these depictions bear little resemblance to historical political configurations in these regions. Of note in the Inner Asian context are Johan Elverskog's (2006) examination of the Mongol view of the Qing imperial project, and David Sneath's (2007) study of aristocratic power and local level state-like processes of administration in steppe communities. Meanwhile work by Geoffrey Samuel (1982, 1993), Melvin Goldstein (1971a, 1989), Georges Dreyfus (1995) and Fernanda Pirie (2005), amongst others, has revealed a far more nuanced picture of traditional Tibetan polities.

Pre-modern Tibet was at its most unified and powerful during its imperial expansionist era under the rule of the first Tibetan emperor, Srongtsen Gampo (618–650), with military expeditions to neighbouring territories, dynastic alliances with Nepal and China, and extensive regional trade networks (Beckwith 1993). However, this relative politico-territorial unity was brought to an abrupt end with the assassination of King Langdarma in 842. Thereafter central authority rapidly fragmented and the region 'reverted to a patchwork of petty states, few of them consisting of more than a couple of valleys' (Samuel 1982: 220). As such, Geoffrey Samuel describes pre-modern Tibet as constituted of an ongoing tension between areas of statelessness and the limited presence of state(like) power. Intriguingly, Samuel suggests that the closest analogies to these 'stateless' societies in Tibet were 'not the Buddhist and Hindu states of South, Southeast and East Asia, but some of the Islamic societies of Central Asia and North Africa' (1982: 215). In both cases low population densities, impeded communication due to the vast and rugged terrain, and low agricultural and pastoral productivity meant that 'centralized political regimes were barely achievable' (*ibid*: 218–219). Under such circumstances it was networks of overlapping allegiances that defined pre-modern Tibet rather than fixed notions of territorial boundaries (Dreyfus 1995).

A defining feature of Tibetan governance structures since the time of the Tibetan Empire has been the prominent role of religion. Indeed, spiritual doctrines have arguably been more influential in shaping Tibetan governance practices, worldviews, collective consciousness and relations with other polities than temporal principles. Buddhism was first brought to Tibet in the seventh century by Srongtsen Gampo and, under the following Tibetan kings, became established as the state religion.[1] Since this period Buddhist principles have informed the Tibetan political system and provided its primary source of legitimacy (Dreyfus 1995). Monastic orders of the four major sects of Tibetan Buddhism (Nyingma, Kagyu, Sakya and Gelug) played an increasingly important role in the political landscape of Tibet, with shifting alliances between aristocratic rulers and religious quasi-states defining the configurations of authority at different scales. Following a prolonged civil war, political and religious authority coalesced from the seventeenth century around the Dalai Lamas, a lineage of religious leaders of the Gelug school of Tibetan Buddhism who are considered to be manifestations of *Chenrezig* (Sanskrit: *Avalokiteśvara*), the bodhisattva of compassion.

Key to consolidating this authority was the fifth Dalai Lama's founding of the Ganden Phodrang government in the Tibetan cultural and commercial centre, Lhasa, in 1642. Named after the Dalai Lama's residence in Drepung monasteries, this government assumed political dominance over central Tibet during the reign of the 'Great Fifth', under whose rule the Ganden Phodrang's performance of statecraft 'drew upon important precedents from Tibetan history as a way of signifying his [the fifth Dalai Lama's] links to Tibet's political and religious heritage' (Mills 2014: 401). The establishment of the Ganden Phodrang also solidified the intertwining of political and spiritual legitimacy, as it saw the establishment of *cho-sid-nyi* (trans.: 'both Dharma and the temporal'), a dual system of religion and politics. This was initially instituted as a diarchal system of a temporal ruler coexisting alongside the spiritual authority of the Dalai Lama, but these roles were combined in 1751 during the reign of the seventh Dalai Lama. Mirroring this dual religious and political leadership were parallel ecclesiastical and secular offices at every level of administration, with government officials recruited from Gelugpa monasteries and Tibetan noble families respectively (Petech 1973; Travers 2011, 2012). Within this system the government functioned with a well-defined hierarchy. The Dalai Lama or, in his absence, a Regent, had ultimate authority over government decisions. Beneath him was a Prime Minister, appointed by the Dalai Lama or Regent, who liaised between them and the lay cabinet (*kashag*, first established in 1751), which oversaw secular matters. A parallel structure of the *chigyab khembo* (ranking monk official) and the *Yigtsang* Office based at the Dalai Lama's residence had authority over religious and monastic matters (Goldstein 1989). Across both these lay and religious administrations the Ganden Phodrang enacted four key governmental functions: 'administering, collecting revenues, storing and redistributing revenues, and deciding cases according to the law' (Dreyfus 1995: 136).

However, whilst the Dalai Lama's rule was, when he attained his majority, '"absolute" in the sense of being beyond challenge in principle' (Mills 2003a: 338), the Ganden Phodrang's authority was nevertheless limited in key respects. Most importantly, the polity of the Dalai Lama and his government comprised only part of the territory subsequently labelled as 'ethnographic Tibet'. Its authority did not reach to any large extent to Amdo and Kham provinces to the north-east and east of Lhasa (see Figure 3.1). Similarly the Tibetan speaking populations of the Himalayan regions 'only occasionally and briefly experienced effective centralized control' (Samuel 1982: 215). For the minority of Tibetans who were ostensibly within the area controlled by the Ganden Phodrang this pre-modern polity's lack of communication infrastructure, policing and military meant that the activities of these 'subjects' were barely monitored or regulated, and the Lhasa regime lacked the capacity to coerce them into following its directives (Mills 2003a). Moreover, the vast majority of Tibetans were hereditarily bound to local estates controlled by monasteries and aristocratic, which continued to enjoy significant autonomy, including tax raising and dispute resolution powers

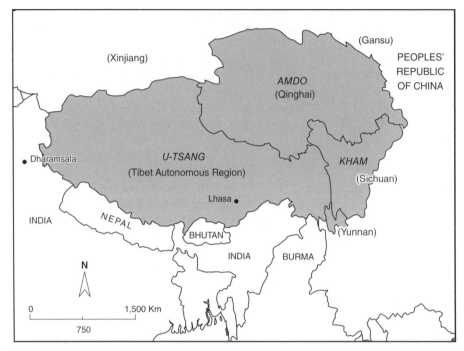

Figure 3.1   Map of Tibet showing traditional Tibetan regions and contemporary Chinese provinces. (Created by A. Allen, 2015.)

(Carrasco 1959; Goldstein 1971a; Pirie 2005). The monasteries in particular were powerful landholders and, as Martin Mills points out, large areas of central Tibet:

> such as those under the Panchen Lama, the massive power blocks of Lhasa's three great monastic centres ... [and] the semi-autonomous Sakya principalities – retained substantial internal autonomy, and on several occasions ... seriously challenged the authority ... of the incumbent Dalai Lama. (2003a: 332; see also Goldstein 1989)

Yet, despite the territorial, legal and infrastructural limitations of the Ganden Phodrang, the Tibetan polity under its rule was nevertheless a remarkably stable system of political and religious governance, and by the eighteenth century had become considerably more bureaucratised (Dreyfus 1995). As such, the modes of power in operation in pre-twentieth-century Tibet lay somewhere *between* a unified state authority under the absolute rule of the Dalai Lamas and David Sneath's notion of a 'headless state' whereby state-like power was constituted by 'horizontal relations between power holders' (2007: 2). When considering the state as a form of social relation (see Chapter 2), then what we can trace here is the central role of symbolic and ritual authority in enabling the Tibetan government to function despite lacking a centralised or comprehensively regulatory system (Samuel 1982, 1993). The drawing parallels to studies of devolved ritual authorities in other historic polities in Asia – for example Tambiah's 'galactic polity' (Tambiah 1977), Geertz's Balinese 'theatre state' (Geertz 1980) and Southall's 'segmentary state' (Southall 1956) – Martin Mills makes a persuasive case for the gradual growth of Lhasa as a 'sacred centre' – being the seat of the Dalai Lamas and site of Buddhism's founding under Srongtsen Gampo – whilst 'symbolic sovereignty was focused at the local level by sacred images and enclosures that replicated the ritual glory of the centre in local monasteries, village temples and household shrines' (2003a: 339). As the chapters that follow discuss, symbolism and, to a lesser extent ritual authority, continue to play a key role in the enactment of governance in the exile manifestation of the Tibetan government.

Prelude II. Tibet and its Neighbours: Contested Narratives of Territory and Authority

The internal structures, modes of governance and administrative limitations of the traditional Tibetan polity provide pointers as to the nature of political organisation in this region prior to the twentieth century, but this tells only part of the story. No polity functions in isolation, and the relations that the Tibetan government forged with neighbouring empires and states expose contested articulations of power, as well as setting the stage for the turbulent events of the mid-twentieth century.

However, before proceeding, any discussion of the relations between Tibet and its neighbours – particularly China – necessitates a note on the politics of Tibetan historiography. As one of the contemporary world's seemingly most intractable conflicts, controversy has pervaded discussions of the legal, territorial and political status of Tibet. The polarised narratives of the two contenders to power in this region can be summarised as follows. Chinese authorities maintain that Tibet has been and remains an 'inseparable part of China' (Wei 1989: 27) since the thirteenth century and, as such, treats its relation with Tibet as one of internal affairs. Beijing's principal claims to Tibet rest upon two historical assertions. First, that the government of the People's Republic of China (PRC) is the rightful inheritor of the territories ruled by the succession of Chinese dynasties, and, second, that since the early medieval period the ethnically Tibetan territories have been subject territories of such imperial rule (People's Republic of China 1992). The counter-narrative posited by the exile Tibetan leadership and their supporters asserts that Tibet was an independent state in the early twentieth century and is thus currently under unlawful Chinese occupation (DIIR 1996; McCorquodale & Orosz 1994). In addition, Tibet scholars argue that, over the centuries, Tibet both extended its influence over neighbouring polities and peoples and, in other periods, itself came under the influence of powerful foreign rulers such as the Mongol Khans, the Manchu Emperors and the British rulers of India (van Walt van Praag 1987). While these various arrangements involved differing degrees of constitutional dependence and independence vis-à-vis China, it is argued that they did not constitute Tibet's subjugation to China, but rather a continued status somewhere between autonomy and independence.

The conflict over Tibet's status is thus 'a conflict over history' (Sperling 2004: ix) and, with no room for complexity within their narratives, also a 'denial of history' by both sides (McGranahan 2010; Shakya 1999: xxii). The consequence has been the effective closing down of 'the space available for scholarship that is not explicitly framed as political advocacy' (Hansen 2003: 9). This has started to change in recent years with a handful of studies examining, side by side, how assertions have been framed in both Chinese and Tibetan sources (e.g. Powers 2004; Sperling 2004; Tuttle 2005). In seeking to 'disentangle the strands of historical argument' (Sperling 2004: 1) regarding Tibet, this scholarship exposes the internal contradictions and inconsistencies in both sides' narratives. However, Tibet studies' late and hesitant embrace of critical theoretical approaches cannot be pinned solely on the politics of knowledge production regarding contemporary Tibet. Anachronistic terminology has been another factor that has contributed to the marginalisation of Tibet within Western-focused scholarship, particularly within studies of geopolitics. Not only is it that 'words such as "theocracy" and "feudalism" simply do not play well to modern Western audiences' (Mills 2003a: 334) but this case exceeds conventional (read 'Western') understandings of statehood, territory, sovereignty and imperial power (Anand 2004). It is ambiguities and contestations surrounding the notions of territory and authority in Tibet's relations with its neighbours that I turn to in this second prelude to the twentieth-century crisis in Tibet.

Just as 'what is Tibet' is a challenging and perplexing question to answer so, equally, is 'where is Tibet?' The area of Tibet has been defined in geographical terms as the globe's largest plateau and the 'roof of the world'; in political terms as the territory coming under the jurisdiction of the Tibetan government; and in ethnic and cultural terms as the wider realm of Tibetan culture, which, pre-dating the rule of the Dalai Lamas, includes the eastern regions of Kham and Amdo as well as regions across the Indian, Nepali and Bhutanese Himalayas. It is this notion of a 'Greater Tibet' – framed as '*Bod Cholkha-sum*' (the three provinces of Tibet, Ü-Tsang, Kham and Amdo, see Figure 3.1) – that underpins exile Tibetan understandings of the territory of Tibet and 'has become deeply embedded in the political culture of the Tibetan diaspora ... crucial in forging unity among diverse refugee groups' (Shakya 1999: 387). However, in promoting this definition to include not only ethno-cultural Tibet but also the territory ruled by the Tibetan government and to which Tibetans have a political claim to, the exiled leadership has been accused of a '"sleight-of-hand" that denies the true boundary of the Ganden Podrang's historical political sovereignty and elides the complex and long-lasting ethnic heterogeneity of the regions under debate' (Mills 2014: 398). This is not to deny the ties of religion, language and culture between Lhasa and the eastern peripheries which have remained relatively strong over the centuries (indeed the present Dalai Lama was born in Amdo), but, as recent scholarship has detailed, many Khampas and Amdowas were 'fiercely contemptuous' of the Lhasa regime (Lopez 1998: 197), even rejecting the descriptor 'Tibetan' (*bod pa*) (Mills 2014: 398). As such, it is notable that the Pan-Tibetan nationalism that is in evidence today is a product partly of common experiences under Chinese occupation since the 1950s and strategically fostered by the exiled elite to encourage unity across what is now a globally dispersed diaspora (discussed in Chapter 6).

In contrast to this Tibetan interpretation of the territory of Tibet, Chinese authorities have consistently refused to recognise any 'Greater Tibet' and use the term 'Tibet' to refer only to the 'Tibetan Autonomous Region' (TAR), a province-level autonomous region of the PRC created in 1965 that corresponds only to Ü-Tsang and part of western Kham. Amdo and remaining regions of Kham were subsumed within the expanded borders of Qinghai, Sichuan, Gansu and Yunnan provinces, with the ethnically Tibetan areas designated as 'Tibetan prefectures' (see Figure 3.1). Notably, just as the Tibetan narrative of *cholkha-sum* denies the messy and partial territorial jurisdiction of the pre-1959 Tibetan government, so the Chinese (and imperial British) delineations of political borders and shrinking of the size of political Tibet deny the existence of forms of state power in pre-1950s Tibet.

In seeking to move beyond these incommensurate interpretations of the territory of Tibet and think more critically about the spatial imaginations and practices enacted in this region is an emerging body of work within Tibetan and Himalayan studies engaging with the notion of Zomia. Denoting a remote hill people, work on Zomia originally focused empirical attention on the South-East Asian Massif (van Schendel 2002),[2] but with James Scott also developing Zomia as part of 'a global history of populations trying to avoid, or having been extruded by, the state' (2009: 328) it is therefore a concept that has the potential to 'travel' (e.g. *Journal of Global History* 2010 5(2)). At first glance there are notable resonances between the Tibetan region and Scott's interpretation of Zomia, from the sparse population and geographical isolation, to the rejection of centralised authority by peripheral pastoralist communities (Pirie 2005). Yet, at the same time, points of divergence are significant. For example, whilst the authority of the Dalai Lama's government may have been relatively weak in comparison to other states, there were nevertheless centres of political, economic and religious authority and, crucially, rather than rejecting the state there were a series of attempts at state-building in the region. However, beyond the empirical insights that Zomia can offer (or not) on the Tibetan case, this notion also provides a useful analytical lens on the role of the state in this region. For example, thinking through the concept of Zomia in light of her research on the Thangmi community, who divide their year between Nepal, Sikkim (India) and Tibet, anthropologist Sara

Shneiderman (2010) argues that Zomia highlights both the ethnic and national fluidity of Himalayan communities and the agency that they have in their engagement with multiple states. Albeit beyond the remit of this study, Zomia may thus provide an alternative and thought-provoking framing for exploring notions of territory in this region,

Returning to contested historiographies of Tibet, what of accounts around authority? Broadly speaking two 'meta-narratives' – religion and imperialism – have framed how each side of this dispute has (retrospectively) narrated political power in this region. The official Chinese narrative of Tibet's incorporation into China has, since the mid-1980s, centred on the then ruling Sakya leaders' submission to the Mongol Yuan dynasty in the thirteenth century. Tibetan historians and politicians, however, not only assert that Chinese claims to Tibet rely on historical relations between Tibet and the *non*-Chinese dynasties of the Mongols (Yuan: 1271–1368) and Manchus (Qing: 1644–1911),[3] but that the relations between Tibet and these empires were based on the traditional Buddhist priest-patron relationship (*chö-yön*; Wylie: *mchod yon*). In essence, Tibetan lamas acted as spiritual guides to successive Mongol and Manchu emperors and, in exchange, received military protection and economic support. Crucially, as these relationships were religious in nature it is argued that they therefore did not constitute Tibet's subjugation to, or unification with, China (Shakabpa 1984). The notion of *chö-yön* has not only become the definitive descriptor of Sino-Tibetan relations in Tibetan historical writing, but its specificity to the Tibetan Buddhist context has been used to reject any discussion of the notion of sovereignty (Sperling 2004; e.g. van Walt van Praag 1987, 2013).

An alternative framing of Tibet's relations with the Qing dynasty is that of feudal vassalage, a relationship of unequal mutual obligation between a feudal tenant and lord. In this rendering, promoted by Chinese writers until the PRC's anti-colonial stance discouraged the use of 'imperial' language in relation to Chinese political history, Tibet is understood as being a tributary polity under subordinate protectorate relations of the more powerful Qing authority. This was a form of relationship common across the Qing empire and beyond, though it declined alongside this dynasty towards the end of the nineteenth century with the arrival of predacious European powers (Bickers 2012). Imperial relations of a different nature were in evidence in the late nineteenth century when Tibet became a focus of attention – and arguably a 'pawn' – in the 'Great Game', with British, Russian and Chinese empires all seeking to have influence over the territory (Kuleshov 1996; McKay 1997). Conscious of potential vulnerabilities on the northern borders of British India, the British attempted to forge allegiances with Tibet, seeking assurance of a friendly buffer state between India and their Russian imperial rivals. Failing to get the reassurance they desired, the then Viceroy of India, Lord Curzon, dispatched a military expedition to Tibet, which, under the leadership of Francis Younghusband, fought its way to Lhasa in 1903–4. The Tibetans were vastly outgunned by the British forces and their state-of-the-art

Maxim machine guns – resulting in thousands of Tibetan deaths – and, with the 13th Dalai Lama having fled for safety to Outer Mongolia, low-level Tibetan officials were forced to sign the Great Britain and Tibet Convention (1904) before the British withdrew. In light of these relations with the Qing and then the British, modern Tibetan history has certainly been shaped by imperialism, but not in a conventional way. Rather, as Carole McGranahan argues, Tibetan imperial experience always had an 'edge' to it, and is perhaps best described as 'off centre' (2010: 173). An important implication of this has been the inadequacy of imperial terminology to describe Tibet's political status. The concept of 'suzerainty' is a prime example.

As Elliott Sperling notes, 'British officials and writers tended to refer consistently to Qing dominance as a form of "suzerainty"' (2004: 6), understood as the extension of protectorate relations over a less powerful polity, but not fully incorporating that polity. This was a term carefully chosen to further British interests in the region as, by acknowledging Tibetan subordination to the Qing dynasty (even when this empire was in decline), the British hoped to block Russian involvement in the region. Just as the Buddhist priest-patron relationship does not easily translate into contemporary Western political lexicons, the same is the case with suzerainty. It is a concept that neither fully corresponds with the traditional Chinese diplomatic theory of vassalage, nor fits with theories of international law in which there is no space for overlapping territorial jurisdictions (Anand 2006). Yet despite – or perhaps because – suzerainty is a term whose vagueness meant it was never full defined or specified, it has had a lasting legacy in the framing of the 'Tibet Question' and 'came to bedevil later interpretations of Sino-Tibetan relations' (Sperling 2004: 6).[4] Indeed, the off-kilter nature of imperial relations that Tibet was subjected to and implicated in, as well as descriptors of political relations that fail to equate with each other, have arguably underpinned the ambiguous status of this case in contemporary international relations: a situation that 'short-changes Tibetan pasts and futures' (McGranahan 2007: 177).

De Facto Statehood Claimed (1911–1949) ...

It was out of the context of the declining Qing Empire and Tibet's compromised position in Great Game politics that the 13th Dalai Lama was, in 1895, able to resume full temporal and spiritual powers and sought to redefine Tibet's status. Trying to pinpoint the emergence of an independent Tibetan state in a more Westphalian guise is, however, a difficult task. Whilst elements are evident in the late nineteenth century, and the Younghusband invasion galvanised the Dalai Lama to assert Tibetan autonomy vis-à-vis the Qing Empire, relations with states beyond its neighbours were not formalised in any meaningful way. The early twentieth century was also a period of political instability in the region. The 13th Dalai Lama was forced to flee Tibet twice: to Mongolia and China in 1904 after the

British invasion, and to India in 1910 when the Qing dispatched a military expedition to Lhasa in an attempt to establish direct rule (Goldstein 1989). A year later, however, the declining Qing dynasty was overthrown in the Xinhai Revolution and, by late 1912, Tibet had expelled the two Manchu *ambans* (imperial representatives) from Lhasa and driven out the last remaining Qing troops. On 13 February 1913 the Dalai Lama issued a proclamation reiterating the priest-patron nature of relations with previous empires and formally stating that Tibet was 'a small religious and independent nation' (cited in Shakabpa 1984: 248). This declaration forms the cornerstone of contemporary claims that Tibet had been a sovereign state,[5] although Tibet scholars are also careful to note that 'this proclamation did not create, but rather described or confirmed, Tibet's status as an independent polity under, inter alia, the formal criteria elaborated by the European law of nations' (Sloane 2014: 60).

Tibet's independence was neither recognised by China, nor formally acknowledged by Britain or any other state. Yet, though denied de jure statehood, Tibet did function independently of China from 1913 until the early 1950s and established a number of attributes of modern statehood. Known as the 'great reformer', the 13th Dalai Lama sought to transform Tibetan politics and society, and his periods in exile in Mongolia and India were particularly instructive in terms of his exposure to secular education systems and modern modes of governance. Back in Tibet, the Dalai Lama established a standing army, introduced economic reforms, issued currency and postage stamps, and extended the reach and sophistication of the civil service bureaucracy. According to Melvin Goldstein, there were 400–500 Tibetan lay and monk officials working for the Ganden Phodrang in the period 1913–951 (Goldstein 1989: 5; Travers 2011).

Yet the increased scale of the Tibetan bureaucracy is not to imply that full territorial control of ethnically Tibetan areas was achieved in this period. Far from it. By the 1950s the Lhasa administration still did not control eastern Kham and most of Amdo. Rather, these regions were ruled by a patchwork of authorities including local princes and chieftains, religious leaders, and Chinese and Hui warlords with whom the central government had varying levels of engagement (Samuel 1993). Yet the lack of defined borders and 'the fact that the central government did not exercise day-to-day control over the whole population' (Goldstein 1971a: 176) does not necessarily imply that Tibet did not exist as a state. Rather, 'it simply means that the nature of the Tibetan system cannot be neatly pigeon holed into a euro-american political framework' (*ibid*).

As evidence of the Lhasa government's 'superordinate authority' Goldstein (1971a: 177) documents its control of the military, regulation of the export and import of key goods (e.g. wool, salt and tea), management of the communications-transportation network, role as a court of last appeal and assertion of its right to enter into binding agreements with foreign nations. It is the latter that has been emphasised by Tibetan writers as providing validation for Tibet's independence. Having met with a series of foreign leaders during his periods in exile the 13th

Dalai Lama was certainly more aware of the importance of international politics than his predecessors, and diplomatic and economic relations were established with neighbouring states of British India, Bhutan, Sikkim and, to a limited extent, Russia and Japan. Most of these states had some sort of diplomatic representatives in Lhasa, and the Tibetan government established a Foreign Office in the early 1940s (Goldstein 1989: 381–385). Tibetan representatives also attended the 1947 Asian Relations Conference in India under their own flag and, later that year, a Tibetan trade mission was dispatched to Europe, North America and India, using Tibetan papers as travel documents (Shakabpa 1984).

The role of foreign powers in attempting to define Tibet's status and its territorial boundaries came to particular prominence with the tripartite British, Tibetan and Chinese conference in the Indian hill-station of Shimla in 1914. Convened by the British, the conference was effectively an effort to settle ongoing disputes over the Sino-Tibetan border, install a politically stable Tibetan administration and thus secure the northern frontiers of British India. The conference began with the three governments on equal footing and the respective plenipotentiaries formally recognising each other's credentials (van Walt van Praag 2013), a point regularly cited by exiled Tibetans as verification of Tibet's independent status. However, after six months of negotiations the conference concluded with only Tibet and Britain signing a draft agreement as the newly founded Republic of China rejected the accord (Addy 1994). Whilst the convention thus has questionable status in international law, its outcomes in terms of border delineation were significant. Under the Shimla Convention, Britain – informed by its lack of desire for both an independent Tibet and full Chinese sovereignty in the region (Strong 1912) – denoted two distinct areas to Tibet: 'Inner Tibet' (Kham and Amdo), which came under direct Chinese sovereignty but over which the Dalai Lama maintained religious authority; and 'Outer Tibet', which came under the 'direct administration' of the Dalai Lama's government, but over which China also maintained a nominal 'suzerainty' (Oberoi 2006: 78).[6] These imperial efforts to delineate Tibet's borders imposed a modernist, statist logic onto a region that, as set out above, had been dominated by overlapping jurisdictions and locally sanctioned boundaries.

Despite Tibet's forays into international politics noted above, these foreign relations never added up to full recognition and, reflecting the isolation that the Tibetan government had cultivated to protect the state from foreign domination, this polity remained reticent to establishing strategic alliances. With regards to domestic affairs, the years after the Shimla Convention were dominated by internal wrangles between factions of the Tibetan elite, and a dispute between the 13th Dalai Lama and the 9th Panchen Lama. As a result, the Dalai Lama's vision for widespread reforms was never fully realised. This stagnation in Tibetan politics was compounded by the fact that, during the interregnum that followed the death of the 13th Dalai Lama in 1933 and preceded the recognition of the 14th Dalai Lama in 1937, the politically conservative monasteries acquired increased powers. In sum, whilst the Tibetan polity incrementally fulfilled a number of criteria of modern statehood during the first decades of the twentieth century, it was not until the threat from the PRC materialised in the early 1950s that the need for international recognition, secured borders and a fully centralised state administration hit home for the Tibetan government. By that stage it was too late.

... De Facto Statehood Lost (1949–1959)

> There was no time to take anything which was not essential with us; we had to be well away from Lhasa before the dawn. The ministers had my Seal of Office, the Seal of the Cabinet, and a few papers which happened to be in the Norbulinka. Most of the state papers were in the Cabinet office or the Potala, and they had to be abandoned.
> Dalai Lama 1997: 158

Following the civil war between the nationalist Kuomintang and the Communist Party of China in the late 1940s, the latter seized control of most of mainland China in 1949, with Chiang Kai-shek's Kuomintang retreating to Taiwan. One of Communist Party Chairman Mao Zedong's first stated goals on proclaiming the establishment of the PRC on 1 October 1949 was the 'liberation' of Tibet from its 'feudal' systems of serfdom and slavery (Powers 2004). Exactly a year later, 40,000 troops of the People's Liberation Army (PLA) entered Chamdo in eastern Tibet. After 12 days the PLA defeated the ill-equipped 8000 strong Tibetan army, and the Tibetan governor of Chamdo capitulated to the Chinese. The *Tsongdu* (Tibetan national assembly) convened an emergency meeting in November 1950 to request the 14th Dalai Lama, then 15 years old, to assume full political authority as Tibetan head of state. With Chinese advances continuing, this time into territory controlled by the Tibetan government, the Dalai Lama appealed to the UN, Britain, the United States and India for assistance, but to no avail.

It is worth pausing a moment to consider postcolonial India's stance on the status of Tibet in this period. As with 'all other countries with which India has inherited treaty relations from His Majesty's Government' (Ministry of Foreign Affairs 1959: 39), in the initial years after Indian independence in 1947 Jawaharlal Nehru followed the British government policy in treating Tibet as a de facto independent state. In line with such a position, whilst declining Lhasa's request for troops in 1949 due to its ill-equipped and fledgling army and significant domestic concerns post-partition, India did not hesitate to deplore China's 'invasion' of Tibet (Mehrotra 2000).[7] However, by 1950 the Indian leadership was increasingly looking at the communist regime in China with admiration and was viewing China both as a powerful neighbour that needed to be placated, and as representing a key postcolonial ally 'destined to lead the emancipation of hitherto dependent Afro-Asian countries' (Chaturvedi 2004: 79). In light of this, while Nehru was advocating the right of self-determination of the peoples of European colonies in Africa and South-East Asia, his position on Tibet and China changed significantly.

No longer recognising Tibet as an autonomous state, India in the early 1950s regarded it as a province of China, a shift that was in direct contravention of the Shimla Convention. As such, Nehru's non-aligned 'neutrality' and strategy of entering into non-aggression pacts with India's immediate neighbours effectively meant an alignment with the Chinese position vis-à-vis Tibet.[8] This stance was confirmed through the signing of the 'Panchsheel Agreement' between China and India on 29 April 1954, whereby India agreed to relinquish its extra-territorial rights in Tibet inherited from the British, and recognised China's sovereignty over Tibet. However, with the Chinese invasion of north-east India in 1962 and the ensuing Indo-Chinese border war, China violated the principles of the Panchsheel Agreement and India's vision of China as a key non-aligned ally was shattered. In response India increased its support for the Tibetan refugees in its territory, including recruiting Tibetans to an elite unit – the 'Special Frontier Force' – within the Indian army, which went on to play a key role in a number of regional confrontations (Wangdu 2013). It is only through increasing economic interactions in recent decades that Sino-Indian diplomatic relations have begun to be 'normalised' (Mehrotra 2000: 65). Yet, despite these shifting Sino-Indian relations, India's position on Tibet has remained unchanged. India neither interferes nor assists the TGiE in its dialogue with Beijing, has declared its non-recognition of the TGiE and abstained from voting on resolutions concerning Tibet at the UN General Assembly in 1959 and 1961.

Returning to Tibet itself, in 1951, with the young Dalai Lama posted for his security to the southern border of Tibet on the advice of the *Kashag*, a delegation of Tibetan officials went to Beijing to negotiate with their Chinese counterparts. There the so-called 'Seventeen-Point Agreement for the Peaceful Liberation of Tibet' was signed between Chinese and Tibetan officials; this affirmed Chinese sovereignty over Tibet, albeit 'with promises of continued autonomy and the preservation of Tibetan religious and social traditions' (Tuttle 2005: 1). This agreement formalised Tibet's incorporation into the PRC and thus ended Tibet's de facto independence. However, the Tibetan ratification of the agreement is a significant point of contention as the delegation exceeded their authority by signing it without approval from the Dalai Lama and the *Kashag* (Kuzmin 2011), and Tibetans and their supporters assert that it was signed under duress and is thus invalid (Dalai Lama 1990; Grunfeld 1987; Shakabpa 1984).

Popular resentment of Chinese rule and the often brutal implementation of communist land reforms in the Tibetan countryside grew during the 1950s. This resentment included armed rebellion in eastern Tibet where Khampas founded the *Chushi Gangdrug* resistance movement (trans. 'Four Rivers Six Ranges', after the Khampa homeland). This guerrilla army later received covert support from the exile Tibetan leadership and the governments of India, Nepal and the United States (including the CIA), and was operational until 1974 when the Dalai Lama directed the soldiers to lay down their weapons (Knaus 2000; McGranahan 2010). In 1959 Tibetan resistance reached central Tibet where there was an open revolt in Lhasa and, beginning on 10 March of that year, daily protests were held in the capital. These were initially motivated by rumours that the Chinese Communists were planning to arrest or abduct the Dalai Lama, and quickly transformed into calls for Tibetan independence. With the Chinese firing artillery shells at the Dalai Lama's residence, plans were made for his escape, in disguise and under the cover of darkness, on 17 March 1959. In the days that followed the revolt was swiftly crushed by the PLA with the death of tens of thousands of Tibetans and the imprisoning of many thousands more (Shakabpa 1984).

It was during the Dalai Lama's journey to India that a temporary government – which prefigured the TGiE – was established through a ritual performance witnessed by His Holiness's small entourage and local villagers. As the Dalai Lama recounted in his autobiography:

> we heard that the Chinese had announced that they had dissolved our government, and that was something on which we could take action. Of course, they had no authenticity, legal or otherwise, to dissolve the government... But now that the announcement had been made, we thought there was some danger that Tibetans in isolated districts might think it had been made with my acquiescence. It seemed to us that the best thing to do was not simply to deny it, but to create a new temporary government; and we decided to do that as soon as we came to Lhuntse Dzong... We held the religious ceremony [on 29 March 1959] to consecrate the founding of the new temporary government. Monks, lay officials, village headmen and many other people joined us on the second floor of the *dzong* [monastery built in a fortress-style], bearing the scriptures and appropriate emblems. I received from the monks the traditional emblems of authority, and the lamas who were present... chanted the enthronement prayers. When the religious ceremony was finished, we went down to the floor below, where my ministers and the local leaders were assembled. A proclamation of the establishment of the temporary government was read out to this assembly and I formally signed copies of it to be sent to various places all over Tibet at that time. (Dalai Lama 1997: 173)

Introducing the Exile Cast and Plot

The Dalai Lama and his coterie of political and religious officials arrived in India on 31 March 1959, and by 1962 around 80,000 Tibetans had made the arduous journey by foot across the high Himalayas into exile. The first Tibetan refugees came from a variety of socio-economic backgrounds including lamas, scholars, merchants, soldiers, farmers and nomads, and the majority hailed from central and southern regions of Tibet. After the peak years of exodus from 1959 to 1961 the borders of Tibet were effectively closed and the political isolation of China meant that there was 'little contact between Tibetans inside Tibet and the refugee community for more than two decades' (Yeh 2007: 652). The second wave of refugees began in the 1980s, with 25,000 Tibetans arriving in India between 1986 and 1996 as a result of reforms in China following the death of Mao and the gradual loosening of travel restrictions imposed on Tibetans. Members of this second exodus – 'newcomers' (*sanjorba*) as they are referred to – left for a variety of reasons including religious persecution, political repression, aggressive sinocisation and a lack of educational opportunities (Hess 2006). An estimated 2000–3000 Tibetans left illegally for India every year from the late 1990s until 2008, with increasing numbers of young children sent by their parents for a Tibetan education in India, young people seeking better education and employment opportunities, and individuals visiting family members in exile (Yeh 2007). However, since the protests across Tibet in March 2008 the flow of Tibetan refugees from Tibet to India via Nepal has all but stopped due to stringent border controls.[9]

Given the truism that the destination of political exile plays a significant role in determining 'the character of the exiles' struggle from abroad' (Shain 1989: xxiv), it is hardly surprising that the host state of India has had a profoundly important influence on exile Tibetan politics. With the arrival of the first wave of Tibetan refugees coinciding with India's early years of independence, not only was it 'inevitable that the lessons of the Indian struggle for independence [were] going to have an impact on those working to liberate Tibet' (Ardley 2002: 119) but, as will be traced in the chapters that follow, the TGiE has in many ways emulated postcolonial India's modes of governance. This includes combining capitalist principles with socialist inspired centralised state planned development, adopting a managerial rationality through a formalised bureaucracy, and promoting the institutionalisation of democracy at a range of scales.

Upon his arrival in India in 1959, the Dalai Lama announced the formation of an exile government. After a brief stay in the north Indian town of Mussoorie, and in consultation with the Indian authorities, the TGiE was established in Dharamsala in April 1960 and departments were quickly formed to oversee the rehabilitation, welfare and education of the refugees, and the preservation of Tibetan culture. On 2 September 1960 a newly created representative body, the 'Commission of Tibetan People's Deputies' was instituted and, a year later, the Dalai Lama tasked his officials with devising a constitution based on democratic principles. Officially adopted on 10 March 1963, and closely following the Indian model, this was the first written constitution in Tibetan history (Sangay 2003). Whilst the final adoption of the 'Constitution of Tibet' was to be left to the entire Tibetan population when reunited in the homeland, the document nevertheless set out guidelines for a future Tibet and a vision for the structure of the TGiE. Combining Western concepts of parliamentary and popular democracy with principles of Tibetan Buddhism, the Constitution was an aspirational document, and it would be several decades before many of its directives would become a working reality in the exile context.

In the 1960s a series of disputes centred around regional divides, and power struggles and financial scandals dogged the exiled community (Norbu 1976), but slowly a new political elite emerged. Over the decades the TGiE was gradually expanded, developed and institutionalised through a series of reforms that reorganised the administration according to democratic principles. A key milestone in this process was reached with the promulgation of the 'Charter of Tibetans in Exile' on 14 June 1991. A revision of the 1963 Constitution, the Charter was specifically designed for the interim exile situation and is effectively the 'supreme law' governing the TGiE. Establishing the three pillars of Tibetan democratic government (executive, legislature and judiciary), the Charter established new institutions and documented the Dalai Lama's vision of the period of exile as a chance to practise democracy in order to implement it in Tibet.

The Dalai Lama remained as head of state and head of government until his decision in March 2011 to retire from political life. As discussed in Chapter 5, this devolution of powers to the elected TGiE leadership effectively ended the rule of the Ganden Phodrang and thereby promoted something of a legitimacy crisis within the community. Exiled Tibetan political leadership is now undertaken by the Prime Minister (the *Kalon Tripa*, 'chief minister', until the title was changed in 2012 to *Sikyong*, 'political leader'), a post that, since 2001, has been directly elected by the diaspora. The *Sikyong* oversees the *Kashag*, the highest executing organ of the TGiE, which makes policy decisions on matters relating to the refugee community and has the main responsibility of trying to 'keep the question of Tibet alive' (Planning Council 1994: Section 1.2.3). *Kalons* (ministers) were initially appointed by the Dalai Lama but since the implementation of the Charter are now elected by the exile parliament.

The first exile legislature, the Commission of Tibetan People's Deputies, saw the abolition of hereditary titles and the traditional system of appointing monk and lay officials to each position. The role of the parliament's then 13 members was largely symbolic and, whilst these deputies were elected from the exile population, their candidacy was based on nomination by the Dalai Lama. It was not until 1975 that candidates put themselves forward for election in the primary rounds (Edin 1992). The current Tibetan Parliament-in-Exile's (TPiE) 44 members have broad legislative powers and responsibilities, which include overseeing the work of TGiE departments, enacting laws, issuing policy decisions and managing the TGiE's finances. The TPiE also liaises with parliaments and NGOs across the world to gain backing for the cause of Tibet, lobbies for support from the Government of India and, during its sessions in March and September, hears public grievances and petitions of Tibetans in exile. The third pillar of the TGiE's democratic structure is the Tibetan Supreme Justice Commission, which, established in 1992, is responsible for framing a judicial code and civil procedures. However, given its position within the state of India, this 'judiciary' is only able to settle civil disputes between exile Tibetans in accordance with arbitrational procedures: all criminal cases are dealt with by the Indian judicial system (see Chapter 4). Finally, three independent statutory commissions reaffirm the democratic status of the TGiE. These are the Election Commission, which conducts and oversees elections to the TPiE, the post of *Sikyong* and Local Assemblies in the settlements; the Public Service Commission, which recruits and trains Tibetan civil servants; and the Audit Commission, which appraises the accounts of TGiE departments and Tibetan public institutions.

The number of government departments has also increased over the decades, with the current administration constituted of seven ministries with distinct portfolios. The Department of Religion and Culture oversees the preservation of Tibetan religious and cultural heritage through assisting with the re-establishment of almost 200 monasteries and nunneries in India, Nepal and Bhutan, producing religious and cultural publications and administering a number of Tibetan cultural institutions. As one of the first departments established in exile, the Department of Education is responsible for overseeing over 70 Tibetan schools in India and Nepal, devising Tibetan curricula and publishing Tibetan textbooks, while the Department of Health runs primary health centres and Tibetan medicine centres in almost all the exile settlements, administers seven hospitals and oversees public health initiatives (see Chapter 6). The Department of Home is responsible for managing the Tibetan settlements, scattered communities, handicraft centres and agricultural cooperatives in India, Nepal and Bhutan (see Chapter 4), and the duties of the Department of Security include ensuring the personal security of the Dalai Lama and providing assistance for Tibetans acquiring and renewing their Indian 'Registration Certificates' (see Chapter 6). Finally, the Department of Finance oversees the TGiE's annual budget, while the Department of Information and International Relations acts as the TGiE's protocol office, disseminating information about Tibet and liaising with international Tibetan Support Groups (see Chapter 7).

With regards to the issue of finances, the funding of governments-in-exile is a challenge that shines a spotlight on the practical and political contexts in which these polities seek to operate. The TGiE has sought to fund its activities through a broad range of sources, in part because of necessity, but also strategically in order to avoid over-reliance on a single donor who might be susceptible to pressure from China. Financial constraints were of particular concern in the first few years of exile given the rapid influx of refugees, most of whom had fled in haste and arrived in India with few resources. The Government of India and a number of international NGOs funded educational and resettlement programmes for the Tibetan refugees, and this support continues to this day. Funding for the exiled administration itself, however, has been harder to secure. The Dalai Lama sold part of the Tibetan state treasury, which had been deposited in Sikkim in 1950, in order to invest in the TGiE's institutions (Office of the Dalai Lama 1969), and around 4% of all donations from foreign agencies is used to cover the TGiE's administration costs.[10] Contributions are also sought from the Tibetan diaspora in the form of a 'voluntary freedom tax' (see Chapter 6). Indeed, it is the TGiE's role in collecting such 'taxes' and in controlling the allocation and distribution of donor aid (McLagan 1996; Roemer 2008) that is crucial to the administration's ability to exercise authority within and foster allegiance from the exile community.[11]

Past to Present: Continuities and Disjunctures

Evident from the overview of the TGiE provided above is a tension between this institution distancing itself from traditional Tibetan political structures in favour of more internationally acceptable democratic modes of governance, and asserting a direct link back to the pre-1959 Government of Tibet in order to cement its claim to be the legitimate representative of the Tibetan nation (see Chapter 5). Continuity with the past also provided much needed stability in the initial years of exile. Solidifying the linear succession of governance in this period was the employment of a number of Ganden Phodrang officials in the exile administration, along with members of the Tibetan aristocracy who had studied in Tibetan missionary schools in northeast India (Roemer 2008). Meanwhile, at the settlement level, early studies by Western and Indian anthropologists documented the perpetuation of traditional structures of local governance and identified this continuity as key to the community's social stability and its adaptation to the Indian physical and social environment (Arakeri 1980; Goldstein 1975; Palakshappa 1978; see Chapter 4).

Perhaps best exemplifying the dynamic between continuity with the pre-1959 Government of Tibet and adaptation to the political environment of exile has been the shifting relationship between religion and politics. In exile, both Buddhism and the figure of the Dalai Lama continue to be central elements, with His Holiness being 'the one institution perhaps which forges together all Tibetans, whether in Tibet or in exile, into one united people' (Office of The Dalai Lama 1969: i; see Chapter 5). This prominence of religion within the exile polity includes 'the doctrines enunciated by the Lord Buddha' (Dalai Lama 1963: v) being enshrined at the core of the 1963 Constitution and 1991 Charter, seats within the TPiE being reserved for representatives of the four sects of Tibetan Buddhism and the traditional faith Bön, and the exile leadership's adoption of nonviolence as a political policy and strategy.[12] In addition, although the monasteries lost their traditional roles as landholders, tax collectors and local political administrators they remain vital institutions in the exile community, with one or more located in each exile settlement. Indeed, Tibetan Buddhism has been preserved and advanced to a far greater extent in India than inside Tibet as, due to the destruction of thousands of Tibetan monasteries during the Chinese Cultural Revolution (1966–1976) and ongoing repression of Buddhist teachings, a number of senior Tibetan lamas have fled to exile and established exile 'institutions' of the major monasteries in Tibet. However, though providing a powerful uniting force for the Tibetan nation, this interweaving of religion and politics has posed substantial challenges, the most significant of which is the fact that the reliance on Buddhist reincarnation to determine leadership succession continues to be a source of political vulnerability (see Chapters 5 and 7; McConnell 2013b). In light of this the Dalai Lama's decision in 2011 to retire from political life and transfer his political authority to elected leaders both separates 'church and state' at the highest level of government, and marks a significant transition in Tibetan politics.

What we see emerging over the decades therefore is a hybrid institution. Selected traditions from the pre-1959 Tibetan government have been preserved and modernised, while new governance practices – drawn both from the 'local' Indian context and from Western models – have been incrementally introduced. The chapters that follow trace these influences on the administration's government structures, cultures of bureaucracy and political ideologies and seek to address the question of what kind of state(ness) the TGiE is rehearsing. More generally, in setting the broader historical and regional scene for the existence of the TGiE, this chapter has sought to contextualise – and in the process trouble – some of the key concepts discussed in Chapter 2. As we have seen in what has been a brief overview of a dozen centuries of Tibetan political history, conceptions of and claims and counter-claims to statehood, territory and political authority underpin this case, and provide an essential backstory to the analysis that follows. However, in order to explore how and why the TGiE functions in the way it does, attention now turns from history to geography: from contested politico-legal narratives of Tibetan pasts to articulations of governance in the exile Tibetan present and in imagined futures.

Endnotes

1   There is also a small Tibetan Muslim population which is believed to have its origins in Kashmir and numbered around 3000 in 1959 (Siddiqui 1991). Some 2000 members of this community now reside in exile, predominantly in Kashmir, and receive some welfare support from the TGiE.
2   Van Schendel (2002) includes parts of south-eastern Tibet in his maps of 'Zomia', whereas Scott (2009) delineates a more limited territorial area that does not include Tibet or the Himalayas.
3   Tibetans also draw attention to the fact that China today does not claim other territories that had been part of the Yuan Empire.
4   This rendition of Tibet's status remained official British policy until October 2008 when the then British Foreign Secretary, David Miliband, asserted that the concept of suzerainty was 'based on the geopolitics of the time' and was thus outdated and misleading (*The Economist* 2008: 64). Since then Britain has recognised Tibet as a part of the PRC.
5   The 14th Dalai Lama made explicit reference to his predecessor's proclamation of independence in his letter to the UN Secretary General on 9 September 1959 where he stated 'I and my Government wish to emphasize that Tibet was a sovereign state

at the time when her territorial integrity was violated by the Chinese Armies in 1950... no power or authority was exercised by the government of China in or over Tibet since the Declaration of Independence by the 13th Dalai Lama in 1912' (Dalai Lama 1997: 218).

6   The demarcation of these zones was set out by the British colonial officer Sir Henry McMahon, and the terms 'inner' and 'outer' Tibet were changed to 'ethnographic' and 'political' Tibet by British political officer Charles Bell in 1920.

7   The Government of India did, however, respond to the Government of Tibet's request for weaponry and supplied rifles, pistols and ammunition in late 1949 (Arpi 2009).

8   It should be noted that although Nehru's position on China/Tibet prevailed, it was not universally supported by his colleagues in the *Lok Sabha* (see TPPRC 2006).

9   The migration of Tibetan refugees from India and Nepal to the West began in the 1960s when the Swiss Red Cross resettled around 1500 Tibetans in Switzerland, a community that continues to thrive today. The success of this experiment led the Dalai Lama to encourage the Canadian and US governments to accept Tibetan refugees and, in 1971–1972, 228 Tibetans arrived in Canada from India and Nepal. One thousand Tibetans from South Asia were granted permanent resident status under the Tibetan U.S. Resettlement Program in the early 1990s. In 1996 these individuals, selected through a TGiE-administered quota system, were permitted to bring their families to the United States, and today the Tibetan population in North America numbers around 15,000.

10 Part of this treasure was invested in a series of light industrial enterprises established by the TGiE in the 1960s but, due to mismanagement, most of these enterprises failed. The funds that were left were invested in a charitable trust established in the Dalai Lama's name in 1964 (Dalai Lama 1990).

11 This bolstering of the political position of the TGiE within the diaspora has striking parallels with the case of the Sahrawi Arab Democratic Republic (SADR), which, in overseeing the distribution of aid across the camps in Algeria acts 'as a buffer between external actors and refugee-citizens, reinforcing Sahrawi political identity as opposed to humanitarian status as refugees' (Farah 2009: 82).

12 Buddhist societies are not necessarily nonviolent, nor is Buddhism itself a nonviolent religion (Jerryson & Juergensmeyer 2010). Nevertheless, based on Buddhist principles of compassion and *ahimsa* – 'to do no harm'-strict adherence to nonviolence has been promoted by the Dalai Lama and TGiE since the 1980s (Garfield 2002; McConnell 2014).

Chapter Four
Rehearsal Spaces: Material and Symbolic Roles of Exile Tibetan Settlements

It's a warm November day and I'm on an Indian government bus leaving the bustle of Mysore in the south Indian state of Karnataka. We travel past Hindu temples, busy markets, palm trees and maize fields. I get off at Bylakuppe, an Indian village strung out along the main highway, and make my way under an archway decorated with the eight auspicious symbols of Tibetan Buddhism. The track passes a State Bank of India – 'Tibetan branch' – Indian shops and houses, a Tibetan tractor repair workshop and an Indian school. Soon I reach a second archway, which marks the formal entrance to 'Lugsum-Samdupling Tibetan settlement'. I have entered a space where Buddhist prayer flags, chortens[1] and monasteries dominate the landscape, the schools, clinics and agricultural cooperatives are all Tibetan-run, and the Indian *sabji walas* (vegetable sellers) and farm labourers file out of the settlement at the end of each working day.

One of the most striking features of the exile Tibetan polity is that it functions with no legal jurisdiction over territory, neither in the homeland of Tibet nor in the host state India. Compounding this is the fact that, within the latter, the TGiE's governmental operations are enacted across a series of highly dispersed and non-contiguous settlement spaces. In this first take on the TGiE's rehearsal of stateness I turn attention to the range of territorialising practices that this administration engages with in its management of over 50 Tibetan settlements in South Asia such as the one described above. In India these range from self-contained agricultural settlements in Karnataka to smaller handicraft settlements

Rehearsing the State: The Political Practices of the Tibetan Government-in-Exile, First Edition. Fiona McConnell.
© 2016 John Wiley & Sons, Ltd. Published 2016 by John Wiley & Sons, Ltd.

in the Indian Himalayas as well as Tibetan communities in major Indian cities. These official Tibetan communities are established on land granted by Indian state governments, and constitute important rehearsal spaces in exile. It is within these settlements that the exiled Tibetan administration appears most state-like and, as I set out in this chapter, it is the everyday management of the settlements, and their important symbolic roles as spaces of nation-building, that is central to the TGiE's rehearsal of governance.

Reflecting the dual aspects of this polity as both state-*like* and state*less*, two sets of debates are engaged with in the discussion that follows. The first is the relationship between territory and state power (Brenner et al. 2003). As discussed in Chapter 2, the concept of territory is widely understood as underpinning the articulation of sovereignty and infrastructural power within the modern state (Giddens 1985; Weber 1978). The use of territory is therefore perceived as 'a key means through which the state is able to coordinate and centralize its operations and, thus, rule its population in a more effective manner' (Jones 2007: 27). This chapter charts how, in keeping with the TGiE's emulation of modes of liberal statecraft, this exiled government aspires to the development of just such a territorialised state bureaucracy. However, in order to start to grasp *how* the TGiE actually enacts authority over space, an approach to questions of territory that does not take sovereignty for granted is needed. Of particular interest here is Joe Painter's writing on territoriality as an aspect of the state's structural effect. Drawing on Timothy Mitchell's notion of the state as the effect of social practices, Painter asserts that territory is 'not an irreducible foundation of state power' but rather should be 'interpreted primarily as an *effect*' (Painter 2010: 1093): as the product of social and material practices and as intrinsically linked to relational networks (see Allen 2003; Massey 2005). Given the TGiE's lack of both contiguous territory and jurisdiction over the spaces in which it does operate, I explore here how this case not only supports such assertions, but is an example where relational networks and interconnected TGiE/Indian state practices are of heightened importance.

The second set of debates comes from a growing body of literature on the space of the refugee camp. The 'camp' in its many guises has become a focus for empirical and theoretical work within political geography in recent years, spurred in part by an engagement with the work of the Italian philosopher Giorgio Agamben on sovereignty, bare life and the state of exception (e.g. Elden 2006; Minca 2005; Turner 2005). At first glance Tibetan settlements in India fail to fit conventional definitions of refugee camps. They are not – or at least are no longer – spaces of major international humanitarian relief operations. Nor are they spaces of insecurity, violence and incubators of insurgency. However, as explored in this chapter, other aspects of refugee camps do have important resonances with this case and can offer revealing insights. These include the constrained temporality and enduring liminality of these spaces (Ramadan 2012), the role of camps as spaces where national identities can be reinforced and attachments to homelands fostered (Malkki 1992, 1995), and these sites functioning as laboratories for future governance (Farah 2009; Mundy 2007).

This chapter will tack between these two bodies of literature, showing how the exile Tibetan case both fits and fails to fit notions of state space and the refugee camp. After setting the context for the establishment of Tibetan settlements in India I turn to the TGiE's rehearsal of state-like governance in these spaces, focusing on its nested hierarchies of administration, the extent to which these spaces have been constructed as sites of governmental experimentation, and the challenges that the exiled administration faces both from within the community and in managing these settlements under Indian jurisdiction. The chapter then turns to the symbolic role of the settlements in the TGiE's promotion of the rehearsal of 'Tibetanness', particularly in terms of forging links to the territory of the homeland and fostering a sense of Tibetan nationalism. The chapter concludes by returning to questions of state space and posits the territorialisation of exile governance.

The Spatiality of 'Tibet' in Exile

As Östen Wahlbeck argues, 'deterritorialisation as a lived experience seems to be intrinsically connected to life in exile' (1998: 4). The shift in the 1990s towards post-structuralist approaches within diaspora and transnationalism literature has seen a rejection of bounded notions of territory in favour of a focus on flows of people, ideas and objects, and the idea of networks (e.g. Brah 1996; Soguk 1996). In light of this, a logical assumption regarding the spatiality of the Tibetan diaspora is that it is both networked and deterritorialised. And in many ways this is the case: transnational connections and practices are extensive and this community is highly mobile (see Hess 2006; Yeh & Lama 2006). Notwithstanding the traumatic dislocation of forced migration from Tibet, this community is, like the multiple diasporas across Asia, also founded upon a tradition of circulation and sojourning (Amrith 2011). Exiled Tibetans travel frequently for religious pilgrimages and teachings, sweater-selling in Indian cities, attending college and visiting relatives. It is commonplace for family members to be scattered across various settlements, monasteries, colleges and working 'abroad' in the West. Such an itinerant existence is summed up by poet and activist Tenzin Tsundue when, in an essay titled *My kind of exile*, he writes,

Ask me where I'm from and I won't have an answer. I feel I never really belonged anywhere, never really had a home. I was born in Manali, but my parents live in Karnataka. Finishing my schooling in two different schools in Himachal Pradesh, my further studies took me to Madras, Ladakh and Mumbai. My sisters are in Varanasi but my brothers are in Dharamsala. (Tsundue 2003: 28)

Whilst this picture of a highly networked and mobile transnational community certainly resonates with conventional understandings of diaspora space, this is far from a smooth 'space of flows' (Castells 1996). Spend any time in Tibetan communities in exile and you become aware of distinct geographies to these networks and mobilities. Perhaps most obvious is the gruelling journey over the Himalayas from Tibet to India, which has been followed by tens of thousands of Tibetans: a route marked by key 'staging posts' including the Tibetan/Nepalese border, the TGiE-run Reception Centres in Kathmandu, Delhi and Dharamsala, and Tibetan schools and monasteries across India to which the refugees are subsequently sent. Mobilities within South Asia also have marked route-ways, hubs and temporal rhythms. Dharamsala, Bangalore and (to a lesser extent these days) Kathmandu, are the key administrative nodes in the community, while religious teachings scheduled according to the Tibetan lunar calendar draw thousands to the *Tsuglhakhang* (main temple) in Dharamsala, Bodh Gaya and the large Tibetan monasteries in south India. Meanwhile, as the main commercial hub, the guesthouses and restaurants of Majnuka Tilla in Delhi fill each season with Tibetan traders en route to buy sweaters from Indian wholesale merchants ('*Tibiti sweater wallahs*') in Ludhiana to then sell on city street-side stalls across India in the winter months (see Lau 2009). Echoing Lisa Malkki's (1992, 1995) work on Burundian Hutu refugees in Tanzania, this community has, over several decades, engaged in a process of *re*territorialisation in South Asia whereby they are defining new spaces for daily life through the establishment of networks, hubs and social practices. Central to this process of reterritorialisation are the series of exile Tibetan settlements.

The Tibetans who followed the Dalai Lama into exile in 1959 were initially accommodated in Indian government-administered transit camps at Missamari in Assam and Buxa in West Bengal. By 1960 there were over 8000 Tibetans at Missamari and almost 1500 monks at Buxa but, despite Indian government assistance, mortality rates were devastatingly high due to the sudden change in climate, disease outbreaks and the arduous journey across the Himalayas. In light of this, the Dalai Lama approached the Indian Ministry of External Affairs with a request to resettle these refugees in cooler places where they might be temporarily employed. In response, the Indian government established 95 construction camps in the Himalayan foothills where 18,000–21,000 Tibetans were employed as road labourers. The remaining monks were relocated to Tibetan monasteries, which were being re-established in Varanasi and in south India. However, the road construction camps were far from a long-term solution. They failed to provide the necessary source of livelihood, conditions were harsh and families were split up. As Rajesh Kharat notes, 'the conditions were so bad that Tibetan refugee officials admitted in 1964 that these workers were worse off than if they had remained in Tibet' (2003: 54). As a result, the Dalai Lama again requested Nehru to rehabilitate the refugees in other parts of India where they could lead a more settled life.

In accommodating this appeal, and in light of the 1962 Sino-Indian border war and recognition that the Tibetans would not soon be returning to their homeland, the Indian government began creating a series of self-contained agricultural settlements. The motivation for India's generosity in providing land for the Tibetan refugees is arguably two-fold. Ideologically, the settlement programme is framed within official postcolonial Indian narratives in terms of a moral duty and humanitarian obligation to assist (e.g. Jawarhalal Nehru's 'Statement re. Situation in Tibet' to the *Lok Sabha*, 27 April 1959, TPPRC 2006: 47–50). The discourse of 'refugee rehabilitation' is recurrent in discussions about the Tibetan population in India in the *Lok Sabha* in the 1960s though, besides distinction made between 'Pakistan refugees' being 'of our country' in contrast to 'foreign' Tibetan refugees, no direct parallels are drawn to the rehabilitation of Partition refugees in the preceding years.[2] Meanwhile, pragmatically, the Indian government believed that the Tibetan agricultural settlements would reduce the economic burden and help India's food needs by bringing unused land under cultivation. The dispersing of Tibetan refugees in relatively remote areas should also be seen in light of Nehru's drive for stability within the post-partition state (Brass 1990).

The state of Mysore (now Karnataka) was the first to reply to Nehru's request for land, and 3000 Tibetans were settled on a 1500-hectare tract of uninhabited forest land on lease at Bylakuppe in 1960. Settlers were initially paid a daily wage for their labour, and villages were established with groups of five persons allocated a 5-acre plot and a one-room tenement. Further land was granted by the Chief Minister of Mysore at Mundgod, and smaller tracts were later made available in Orissa, Maharashtra, Ladakh, Arunachal Pradesh, Uttar Pradesh, Madhya Pradesh, Sikkim, West Bengal and Himachal Pradesh.[3] By the end of the first decade in exile, 30,000 refugees had been resettled into these designated sites and, by the early 1980s, all existing settlements were established and, in most cases, were at capacity. As a result, few Tibetans who have come to India in recent decades have moved to the settlements, with the exception of those joining monasteries and residential schools. This has led to a growing 'unsettled' Tibetan population, a high proportion of whom reside in a series of scattered Tibetan communities in the Himalayan states and in major Indian cities. These are areas where the TGiE does not administer any land or housing, but where TGiE-run 'Welfare Offices' manage Tibetan schools, clinics and cultural centres, liaise with local Indian authorities, and resolve minor civil disputes.

The Tibetan settlements and scattered communities in India are thus non-contiguous and highly dispersed spaces extending from the remote north-east to the tropical forests of southern India, the high-altitude mountains of Ladakh to the banks of the Yamuna River in north Delhi. In order to reflect this diversity six settlements and scattered communities were the focus of this research (Figure 1).

As the first agricultural settlement to be established, Lugsum-Samdupling at Bylakuppe, Karnataka, remains one of the largest and most populous with

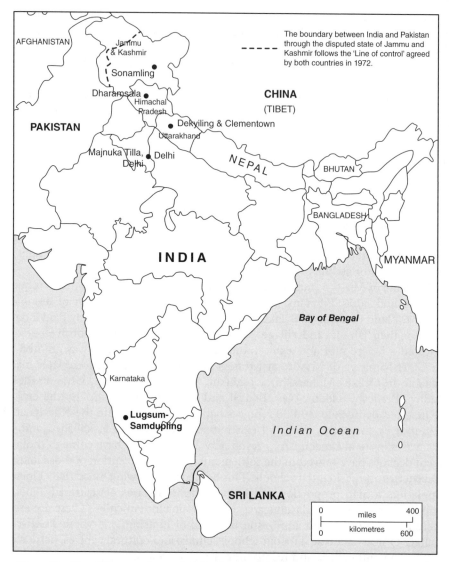

Figure 4.1 Map of Tibetan settlements in India where research was undertaken. (Compiled by E. Oliver.)

over 10,000 Tibetans residing in 3500 acres.[4] Seen as a pioneer rehabilitation project, Lugsum-Samdupling has served as a model for the planning of later settlements, and continues to be viewed as one of the most developed and successful exile Tibetan communities. The settlement consists of seven dispersed villages, or 'camps', each accommodating around 30 families. The four sects of Tibetan Buddhism have also rebuilt monasteries next to the settlement, with the largest,

Sera, housing almost 3000 monks. In stark contrast to the infrastructure, facilities and self-sufficiency of Lugsum-Samdupling, Sonamling agricultural settlement in Ladakh faces a number of challenges due to its high-altitude environment, remoteness and poor infrastructure. Established in 1969 on 522 acres near the Ladakhi capital of Leh, the initial population of 617 were mainly nomads from western Tibet. Today the population of over 5000 reside in 12 camps and, in order to supplement agricultural returns, the settlers are also engaged in trading, casual labouring and seasonal sweater-selling.

Research was also undertaken in two distinctive settlements around Dehradun, a prosperous Indian town in Uttarkhand, several hours north of Delhi. Dekyiling Settlement was established in the early 1980s for the rehabilitation of Tibetan refugees from Bhutan,[5] 'unsettled' Tibetans in the district of Dehradun and retired TGiE officials. The land allocation of 31 acres was intended only for housing, with employment based on handicrafts manufactured on site, seasonal sweater-selling and the service sector in Dehradun. The population has increased from 720 in 1981 to around 2000 today, and the settlement has a dispensary, school and monastery. Across town from Dekyiling, the Tibetan community at Clementown was established not by the TGiE but by a group of Tibetans from Amdo. The settlement has a similar population size to Dekyiling and continues to be run by the settlers themselves, with their 'Clementown Settlement Society' managing the health clinic, school and carpet-making centre. Finally, considerable time was spent in two scattered Tibetan communities. Dharamsala, a hill-station in Himachal Pradesh, has been the 'capital' of the exile community since 1960 and is home to the TGiE, Dalai Lama and numerous Tibetan NGOs, media outlets, and cultural and religious institutions.[6] Frequently referred to as 'Dhasa' (an amalgamation of 'Dharamsala' and 'Lhasa') by its Tibetan residents, Dharamsala has become the geographical centre and base of power for the exiled community, providing a key focal point for pilgrimage, religious teachings and political activism. Lastly, research was also undertaken in Majnuka Tilla, a Tibetan colony established in the early 1960s in north Delhi. This is the key commercial and transport focal point for the exile community, with refugees, pilgrims, traders and students passing through en route to Dharamsala, Nepal (and from there Tibet), the settlements and monasteries in south India and the West. As senior civil servant, Thupten Samphel it in *Tibetan Bulletin*:

[while] Dharamsala is considered the heart of the Tibet world ... it is MT [Majnuka Tilla] that constitutes the commercial centre of the exile community. It is the hub of Tibetan commerce and spreads its limited prosperity along its many spokes to other Tibetan communities in all four directions of the subcontinent and beyond. (2006, 10(5): 27)

Complex histories of negotiations with the Indian government at central and state levels alongside their heterogeneous locations, climates and environments means that the Tibetan settlements in India vary considerably in size, land use and economic activity. Yet, despite this, each settlement has a similar administrative set-up and there is a significant degree of uniformity of TGiE governance across these spaces.

Rehearsing State-like Governance

Although never formally recognising the TGiE as a government, from the early days of exile the Indian authorities welcomed the delegation of responsibility for the Tibetan settlements offered by the exiled administration, granting it a virtual monopoly to represent the refugees (Goldstein 1975). Thus, whilst in the first few years the Indian government was ultimately in control of the settlements – initially through the Ministry for External Affairs and then the Ministry of Rehabilitation and Central Relief Committee – over time authority gradually transferred to the TGiE. It is this degree of Tibetan autonomy within the settlement spaces that enables them to be constructed as sites of rehearsal of – and experimentation with – different modes of governance. In addition, it is through the TGiE's development of an increasingly sophisticated set of mechanisms to circumvent the challenges raised by the dispersed spatiality of 'its' territories that the administration constructs itself as a government. In exploring strategies of core-periphery government structures, uniformity of service provision and nested jurisdictions, two questions posed by Ferguson and Gupta in their study of the neoliberal state in India are pertinent: 'How is it that people come to experience the state as an entity with certain spatial characteristics and properties? [and] Through what images, metaphors, and representational practices does the state come to be understood as a concrete, overarching, spatially encompassing reality?' (2002: 981)

From the early days of exile, the re-established Tibetan government sought to represent and actively manage its refugees, and did so in ways that were explicitly territorial. For example, posting its representatives to the numerous transit camps and early settlements enabled the TGiE to keep 'in touch with the refugees in order to … gauge their needs and provide for them' (Office of the Dalai Lama 1969: 173). In addition, the exiled administration quickly established itself as the central authority within the exile community. As Melvin Goldstein explained:

> the ready availability of a core of highly experienced and competent governmental administrators provided the Tibetans a ready-made organisation through which resources could be effectively aggregated and policy decided on and implemented … The Dalai Lama's Government offered the scattered Tibetan refugees a centralised and efficient organisation which could integrate and represent their needs. (Goldstein 1975: 20)

The consolidation of this centralised authority is key to how the TGiE has attempted to govern its non-contiguous settlement within India and is a strategy that has clear resonances with assertions that territorialised coordination of operations is an important mode through which the state is able to govern its population (Mann 1984; Sack 1986).

A key aspect of centralised power is its relationship to the periphery, and a number of interviewees described relations between the Dharamsala-based TGiE and the settlements as one of a central government and series of local governments, characterised in terms of hierarchies, communication flows and, indeed, the centre's misunderstanding of local issues. Strong connections between Dharamsala and the settlements are maintained by annual meetings of Settlement Officers in Dharamsala and regular visits of Tibetan *chitues* (MPs) and TGiE officials to monitor progress on projects, discuss the implementation of TGiE policies and hear grievances. Illustrating a familiar core-periphery dynamic – albeit one with hundreds of miles of Indian territory between the Tibetan 'capital' and its outposts – the distance between the settlements and Dharamsala was evident both materially and in the imagination. Those who are posted to settlement offices, schools and clinics in remote areas like Ladakh often described such stints in terms of going above and beyond service duty for the government, community and wider Tibetan cause. Meanwhile, visiting the remote settlements in Arunachal Pradesh was seen as a considerable inconvenience by *chitues*, who had 'drawn the short straw' by being sent on a delegation there. Reflecting on how MPs approached such visits one younger parliamentarian explained:

> so many of the older *chitues* they go to the settlements and preach and scold the people. They lecture them about how to think and behave ... The settlement people are viewed as uneducated, as need [*sic*] to be told what to do by the government (February 2007).

In a revealing transposing of geographical imaginations, a retired official even reflected that Dharamsala's relationship to the remote settlements in northeast India and in Ladakh was the exile equivalent to the pre-1959 Tibetan government trying to keep tabs on remote and unruly areas of Kham.

In practical terms this core-periphery model of governance is played out through a hierarchy of administrative authority that the TGiE has developed across different territorial scales. The first tier of administration below the TGiE departments in Dharamsala are 'Chief Representative Offices' (CROs) located in Karnataka, Ladakh and Uttarakhand. These offices oversee the implementation of TGiE policies in their cluster of settlements, distribute project funding and liaise between the respective Indian state governments, the Tibetan communities and the TGiE. Below these regional-level CROs, each settlement has a Settlement Officer and each sizeable scattered community a Welfare Officer. These officials are charged with overall control of the settlement, sitting on school, monastery, hospital and cooperative society committees and acting on behalf of settlers in their dealings with Indian authorities. The TGiE has been trying to encourage settlements to elect their own Settlement Officers and CROs from among their communities as a means of promoting 'good governance', but only a handful set-tlements have taken up this initiative to date. Instead, these officials are usually appointed by the TGiE, spend three years in any given settlement and are there-fore not 'settlers' per se and have no allocated land. Underpinning this preference for officials sent by Dharamsala is a widely held perception that these individuals are direct representatives of the Dalai Lama and are thus holders of authority and worthy of respect.

Where the TGiE has been more successful in reinforcing 'active and demo-cratic grassroots participation both in decision-making and the day-to-day func-tioning of the Settlements' (Planning Council 1994: Section 1.2.7) is through the institutionalisation of Local Assemblies. These assemblies are elected from within the settlement community for three-year terms, and function as the legislature to the executive role of the Settlement Office. They are therefore responsible for passing and auditing the local budget and drafting rules and regulations for the settlement, from restrictions on liquor and gambling, to the terms of communal labour for infrastructure repairs, and the organization of *pujas* (prayers). Follow-ing the central/local government relationship discussed above, the establishment of Local Assemblies can thus be seen as augmenting the settlements' function as municipalities with their own local government structures and democratically elected leadership.

The process of administrative integration across the settlements is significant, with the extension of the TGiE's presence and governance over these disparate territories creating a remarkably cohesive bureaucratic landscape. A similar pic-ture emerges with regards to welfare provision. Although Tibetans have little con-trol over the provision of utilities in the settlements, with electricity and telephone lines provided by Indian utilities companies, and water and access roads provided by state governments, the TGiE does manage physical infrastructure projects such as flood protection and the construction of buildings, and runs basic welfare ser-vices. In terms of the latter, senior TGiE officials were insistent on their *duty* to provide integrated and standardised provision of, and access to, health, education and welfare support across the settlements in India:

> wherever possible we are trying to provide ... at least in the case of welfare activities and education it's almost standard everywhere. They may not have a school there in that locality but they can send their children to the residential schools. In the welfare activities we almost provide the same for everywhere ... standard like that. Even the medicines are the same cost in each clinic! (Additional Secretary, Department of Home, November 2007)

> we have yearly meetings of the Settlement Officers to discuss the major policies in the settlements and to make unanimous rules ... so the same thing happening in all

settlements. Standardised so there's no competition, no-one is saying one settlement is getting priority. Of course there are different personalities so some policies are implemented differently but we try to make equality. (Secretary, CRO Bangalore, November 2007)

Such deliberately integrative policies are key to the TGiE's attempts to institutionalise its authority over the exile population. Moreover, whilst acknowledging that one of the great conceits (or deceits) of most state theory is 'the suggestion that the writ of London or New Delhi or Islamabad reaches without interruption... to the trenches that are at the bottom of the state hierarchy' (Corbridge et al. 2005: 35), it is TGiE's *aspirations* to such flows of authority that is particularly revealing of its stateness. With regards to managing its dispersed settlements, the TGiE employs what appears to be a conventionally state-like top-down articulation of power across its un-state-like territory. Importantly, as 'the production of legitimacy... requires constant enactment of the state as a symbolic centre of society, the source of governance, the arbiter of conflicts, the site of authorisation' (Hansen 2001: 225), this centralisation of power thereby establishes the TGiE as a locus of authority.

Yet it is also apparent that the existence of defined Tibetan spaces in India in many ways enables the functioning of the TGiE as a government. Territory still matters. For a start, TGiE's scalar hierarchies across its bounded spaces in exile echo Agnew's description of territoriality as working 'through territorial division of space, boundary control, and the hierarchical dissemination of authoritative commands' (Agnew 2005: 442). As an example, when describing the spatial reach of their responsibilities, staff in CROs frequently spoke of various settlements, scattered communities, schools and monasteries as coming under their 'jurisdiction'. If the state is spatially understood as 'a nested hierarchy of discrete, enclosed jurisdictional spaces' (Cox 1998: 1), then this discursive construction of bureaucratic jurisdictions – the TGiE's lack of official sovereignty over these spaces meaning that is far from *legal* jurisdiction – can be seen as a conventionally state-like articulation of power over space. Crucially, and discussed in more detail in Chapter 6, the network of territorial settlements functions as a spatial form of regulation, echoing Luke's observation that:

> Territories are... highly politicized formations inasmuch as they structure governmentality; they arrange people with vital systems of things as individuals and collectives, giving access to places used for getting security benefits, health services, identity codes and infrastructural goods. (1996: 503)

In the case of the TGiE, such territorial governance of the diasporic population in India includes monitoring people's movement, recording demographic data, collecting 'taxes', renewing identity documents and facilitating the provision of

Tibetan-run welfare services. We can therefore see a striving towards the two principles of 'state spatialization' that Ferguson and Gupta (2002) argue help to naturalise a state's authority and secure its legitimacy. These are 'verticality' – of the state being above society – and the state 'encompassing' its localities. It is through the bureaucratic practices of TGiE's nested hierarchies of administration, the construction of 'jurisdictions' and the integration and standardisation of services that such state-like spatialisations are produced in what are unlikely circumstances.

Another element in this production of scalar hierarchies is the third pillar of democracy – the judiciary – which is slowly being rolled out across 'national', 'regional' and 'local' levels. As the Secretary of the Supreme Justice Commission explained:

> there are three layers to our judiciary, so first is Supreme Justice Commission, then we have Circuit Justice Commission and then Local Justice Commissions. For the Circuit Justice Commissions which is the regional level there is provision in the Charter for five but ... so far we ... it has not yet been implemented. For the Local Justice Commissions these are for every settlement but only two are so far fully-fledged and so for most settlements the Settlement Officer he is also the Local Justice Commissioner (March 2006).

Given the TGiE's existence within the sovereign state of India, its judiciary has very limited powers, only able to hear civil disputes from within the Tibetan community, and, without powers of detention, is extremely restricted in the punishments it can impose let alone enforce (Duska 2008; see also Chapter 5). In light of this, and without prior experience of secular judicial practices in pre-1959 Tibet (French 1995), there has been a distinct lack of engagement with the different levels of the Justice Commission. For example, whilst Dekyiling and Lugsum-Samdupling have multi-storey buildings for their 'Local Justice Commissions', when I visited only a handful of cases were pending at each and the buildings remained unused except for occasional school visits. As a farmer in Lugsum-Samdupling told me:

> with our justice commission, there are some staff, but not so much work... people here we joke it's the easiest job! No one goes to the justice commission if they have a problem. I think our people are scared of the legal system – they always sort out arguments among themselves, not in the court (November 2007).

In the case of both the Local Justice Commissions and Local Assemblies these institutions thus often fail to live up to their potential, with confusion regarding their remit and function and a feeling amongst some settlers that they provide little more than an extra title for members of the settlement elite (see Frechette, 1997, for disagreements over the role of Local Assemblies in Kathmandu). In contrast, it is often non-TGiE leadership structures within the settlements that are viewed as the preferred point of contact for everyday matters. Central to this is the figure of the *gyapon*. Meaning 'representative of 100', the *gyabon*, or camp leader, is a 'local' Tibetan who acts as an intermediary between the Settlement Office and residents. It is often the *gyabon* rather than the Local Justice Commissioner who is sought to resolve minor disputes, and the *gyabon* rather than Local Assembly members who is consulted over general management issues. Reporting to each *gyapon* are a series of *chupons*, block leaders who represent around ten households and who are responsible for the organisation of community meetings, projects and festivals. Anthropologists researching the early years of the settlements noted how the continuation of these traditional decimal leadership structures from Tibet was key to the relatively smooth transition to life in exile (Palakshappa 1978). Their continuation today indicates both a strong collective memory within the settlement communities, and the resilience of traditional administrative systems alongside Dharamsala's drive to introduce the institutions of liberal democracy.

However, despite the under-use and limited powers of the Justice Commissions and Local Assemblies, both government officials and settlement residents were keen to stress how the existence of these bodies remained important for two reasons. Firstly, they are symbolically significant in completing the tripartite structure of democratic governance (legislative, executive and judiciary) and thereby validating the TGiE as a 'democratic government'. Secondly, they fulfil important aspirational and pedagogical roles. An official at the Supreme Justice Commission summed it up when he explained:

> in the action or practical sense our Justice Commissions they have not so much power, but they are established to complete the three pillars of our government, and for us to practice how we are going to use it in Tibet. So people are familiar with the systems. It is for our experience that we have these commissions established (March 2006).

Local Assemblies and Local Justice Commissions are only two of a number of settlement schemes whereby the exile government is seeking to try out and rehearse different modes of governance. In recent years, a range of initiatives have been promoted under narratives based around Gandhi's idea of *satyagraha*. With its antecedents in the Dalai Lama's vision for future Tibet as a 'zone of *ahimsa*' (TGiE 1992), these schemes were pioneered by former *Kalon Tripa* Samdhong Rinpoche who, in a statement to the press after taking office in 2001, explained:

> I believe we Tibetans need to establish a non-violent society to serve as a model for the rest of the world. In order to do this, we should first develop a culture of *ahimsa* in our exile communities ... Tibetan settlements in India are ideal places for undertaking this experiment. (Tibetan Bulletin 2001: 25)

Seeking to implement this vision of 'Gandhi-like ashrams where traditional culture and [our] Tibetan ways of life are preserved and protected from outside influences' (Tibetan journalist, Dharamsala, April 2007) has entailed the promotion of organic farming, solar and wind energy projects, and soil and water conservation. Attention has focused on Mundgod settlement in Karnataka with the aim of establishing it as a model for these agricultural innovations, which can then be replicated in the other settlements and 'transplanted to a future Tibet' (Agricultural Officer, Lugsum-Samdupling, November 2007).

Such policies are thus presented as central to the wider TGiE project of using the period in exile as an opportunity to experiment with and train the diaspora in political systems and resource use that could be transferred to the homeland should the desired 'return' materialise. As such, the settlements are framed as sites of experimentation where a certain way of life can be trialled and cultivated and a range of practices of statecraft rehearsed. This analogy of the settlements as laboratories of Tibetan state-making has striking parallels with two other cases of exiled polities: the Palestinians in Lebanon (Hanafi 2008; Rougier 2007) and the Sahrawis in Algeria. Writing about the Polisario-run camps in Algeria, Randa Farah describes these sites as 'incubators of new social and political institutions transportable to national territory upon repatriation' (2009: 80, 76). In a similar vein Jacob Mundy argues that, through providing 'a chance to practise the forms of governance and social organisation they would establish following independence' these camps 'became a microcosm, a pre-figurative lived model, of what an independent Western Sahara would, and still could, look like' (2007: 275; see also San Martin 2010). What we have in the Sahrawi and Tibetan cases, then, is a particular articulation of the role and purpose of 'refugee camps'. These camps and settlements have been – and to some extent continue to be – spaces of humanitarian assistance where 'a technology of "care and control"' (Malkki 1992: 34) has been cultivated. But they are also politicised spaces where the production of visions of a national future can confer legitimacy on the respective exiled administration (cf. Feldman 2008).

The TGiE's settlement initiatives and programmes can thus be read as practical means through which imagined futures are made present. However, such visions of the settlements are neither universally shared nor universally supported by those in the diaspora. As noted above, the 'new' legislative and judicial institutions are often seen by settlers as of limited practical use, and some poorer farmers have been resistant to a shift to organic practices, which in some cases reduce yields in the first few years. Of more pressing concern is the fact that many young, educated Tibetans simply do not see their future within the settlements. They are increasingly rejecting these closed, tight-knit Tibetan communities in favour of employment opportunities in Indian cities and emigration to the West. There is certainly a case to be made that the focus of the settlements as sites of rehearsing for the future has meant that practicalities of life in the present have been somewhat sidelined. Responding to what has been framed as a pending crisis within the settlements, the TGiE has shifted its stance in recent years to focusing on the settlements as spaces of opportunity of and for the *present*, a shift that is in line with statements from the *Sikyong* that the issue of Tibet may remain unresolved for another 50 years (Tibet.net 2012). Alongside initiatives to promote economic self-sufficiency in the settlements through small business loans, vocational training and diversifying the activities of settlement cooperatives (TechnoServe 2010), senior officials in the TGiE are also beginning to concede that the settlement system as it currently stands may not be sustainable in the long term. As I write it is still early days in formulating alternative strategies for the settlements, but there are discussions of schemes to merge smaller settlements and to establish a 'sister settlement' (sister *shichak*) programme whereby Tibetan communities in the West are paired with settlements in South Asia.

Limits to the Settlements as 'State Spaces'

The settlements are thus also increasingly sites of vulnerability within the community. By focusing on the administrative systems underpinning the management of these spaces, the discussion above has perhaps painted too cohesive an image of the settlements, and certainly one that tells only part of the story. To balance this narrative it is necessary to widen the focus to look both at how the settlements function within the sovereign space of India and internal challenges to the TGiE's territorial authority. With regards to the former, on the one hand, the fact that the settlements are enclosed extraterritorial spaces *within* but not really *of* the Indian state in many ways facilitates their role as autonomous spaces of experimentation and training. The settlements are predominantly located in remote regions, a geographic distancing redolent of refugee camps more generally whereby they are deemed less accessible to national political elites and their inhabitants' political protests less visible.[7] This sanctioning of Tibetan settlements separated by social and cultural boundaries from the host society has not only facilitated the TGiE's nation-building project in exile (as discussed below), but is also in Indian interests. It corroborates and enacts India's liberal 'non-assimilative' or 'unity in diversity' framework (Goldstein 1978) whereby Tibetans are broadly regarded as a pseudo caste community that, within the framework of India's Hindu caste hierarchy, can maintain social and cultural practices (Norbu 2004).

On the other hand, the fact that India has not ceded jurisdiction over these spaces poses important limitations to the rehearsal project. While the settlements appear to be autonomous Tibetan spaces where Tibetans are ostensibly free to continue their way of life, at the end of the day this remains Indian territory, where Indian law applies and where the Indian government is the ultimate authority. Settlement land is on lease from respective Indian states to the TGiE, and settlers pay land revenue at rates fixed by each regional administration. However, no such collective land grant exists in the scattered communities, and even the exile government itself is a tenant on land rented for its headquarters from the state of Himachal Pradesh.[8] With the Tibetan administration and settlers being subject to Indian jurisdiction, the Indian police thus have authority for law and order within the settlements, and any criminal matters are dealt with by the Indian courts rather than Tibetan judiciary. Such a division of legal responsibilities between the two administrations, with TGiE dealing with civil disputes and Indian authorities with criminal cases, is in many ways mutually beneficial. Tibetans benefit from India's legal pluralism whereby minority communities are granted the freedom to regulate community issues according to customary norms, and the devolution of dispute resolution to the TGiE means that the already overstretched Indian justice system is not burdened by Tibetan civil grievances.

Whilst this division of juridical responsibilities is clear-cut at the national level, on the ground it is more ambiguous. In general, settlement-level Tibetan officials ensure that their communities understand and comply with Indian laws – such as those governing residence certificates and business licences – and act as intermediaries between Tibetan settlers and a range of Indian officials. So, for example, as the Agricultural Officer at Lugsum-Samdupling explained, if Indian cattle wander onto Tibetan land, Tibetan farmers go to their Settlement Officer, who contacts the local Indian officials, who in turn approach the offending local farmers. Similarly, if local Indians have grievances with Tibetans they can, in person or through their *panchayat* (village council), approach the Settlement Office, which will then take up the issue with the settler concerned. Such 'working relationships' are also evident in relation to the issue of defining and policing settlement boundaries.

The delineation of settlement spaces varies considerably across different regions. For instance, given the ethnic similarities between Tibetans and Ladakhis, the boundaries of Sonamling settlement are barely demarcated. Apart from the Tibetan 'branded' schools and clinics, Tibetan and Ladakhi villages are often indistinguishable, and Sonamling itself is not a contiguous settlement, with land between the camps belonging to Ladakhi farmers and Indian military units. In the more self-contained settlements around Dehradun these spaces are conspicuous in the Indian landscape by entrance ways marked by decorative archways – visible 'border posts' – and prayer flags adorning the roofs of settlement buildings. With the larger landholdings the boundaries of Lugsum-Samdupling settlement are somewhat 'fuzzier' but are nevertheless recognised both by Tibetan settlers and local Indians. For example, in areas where the settlement bounds Indian land, there is the striking sight each day at dusk of Indian labourers filing out of the Tibetan area to their villages beyond the settlement boundary. Thus, unlike conventional refugee camps there are no checkpoints or guarded entrances to Tibetan settlements. However, in certain circumstances boundary enforcement practices are employed and the border between two communities of different legal standing sharpens into focus. From the Indian side, local officials have the authority to grant individual Tibetans the right to leave and enter the settlements. Far from daily monitoring, this is bureaucratic control of residency with, for instance, families leaving settlements such as Dekyiling for their winter sweater-selling needing to seek 'departure permission' from the local Assistant Commissioner, and then register their return in the settlement at the end of the season. The extent to which this is enforced varies considerably across settlements and over time.

Yet this is not a one-way enforcement of settlement boundaries. On the TGiE side, although Indians are free to come and go in the settlements, Tibetan officials do monitor and police what goes on in these spaces. For example, whilst chatting to a Tibetan tea-stall owner in Dekyiling I was approached by two Indian charity workers soliciting donations. Before I could reply, the stall owner had demanded to see a 'permission letter from the Settlement Office' from the men. Unable to produce such a document, and with the threat that they would be marched to the TGiE office to explain themselves, they quickly left and headed back to the main road. Though a seemingly mundane occurrence, the accretion of decades' worth of such everyday bordering processes, alongside TGiE officials' management of activities within the settlement spaces, implies a significant degree of de facto authority regarding practices of exclusion, differentiation and identification on the part of the Tibetan administration (cf. Rajaram & Soguk 2006). In sum, the relative autonomy of the Tibetan settlements within Indian territory and the relational networks between Indian and exile Tibetan 'state' actors points to a nesting of polities. As Ferguson and Mansbach (1996) outline, such embedding of one polity within another therefore encourages a focus on the vertical dimensions of political life, which, unlike attention paid to horizontal dimensions, brings into view relations between polities that lie outside of so-called Westphalian norms.

In addition to the legal limitations of functioning within the state of India, the framing of the settlements as sites of rehearsal, and the TGiE's governmental reach more generally, also faces challenges from within the exiled community. A potential contender to the exile government's authority within the settlements comes from the monasteries.[9] Although they have lost many of the important economic and political roles they had in Tibet, the monasteries remain vital institutions in the exile community and are often well funded by international networks of practitioners and supporters. This is particularly the case in the large settlements in south India where many of the sects of Tibetan Buddhism have re-established their main monasteries in exile.[10] These institutions fund infrastructure projects within the settlements and, in the case of Sera monastery at Bylakuppe, runs its own 'monk village' complete with agricultural land, housing, shops and clinics. However, though autonomous in many regards, the monasteries do ultimately come under the jurisdiction of the TGiE-run CROs and Settlement Offices, which monitor the arrival and departure of monks in their region or settlement and, in principle, have the final say on decision-making in these spaces.

An additional challenge to the TGiE's monopoly over the settlements comes from alternative forms of lay leadership, as is evident in Samyeling colony at Majnuka Tilla in north Delhi.[11] Established in 1960 by refugees who had initially been accommodated in Arunachal Pradesh and Nepal, Majnuka Tilla was never fully recognised as a Tibetan settlement by the TGiE and was for many years run by the refugees themselves. As the TGiE-appointed Secretary at the Welfare Officer explained:

> unlike other settlements established by our government where a letter was sent by Government of India to Tibetan government saying here is the land, you are in charge, you organise the settlement, here you see the people began the colony them- selves and so it was never registered with our government – and so it doesn't enjoy privileges from Tibetan government as other settlements (April 2006).

However, with increasing fiscal capacity and a desire to unite the exile com- munity, in 1983 the TGiE 'established a Welfare Office for the Majnuka Tilla people and appointed here a government officer to look after the settlers' wellbe- ing' (older female resident of Majnuka Tilla, June 2007). Given this history, the colony has two distinct leadership structures – the original 'local' leadership and the more recent TGiE leadership – with a clear differentiation of roles and respon- sibilities. The TGiE-run Welfare Office manages the Tibetan government institu- tions within the colony (a basic medical centre, branch of *Men-tse-khang* Tibetan medicine clinic and primary school), administers the Department of Home wel- fare stipends, and acts as an intermediary between the residents, the TGiE and local Indian authorities. In contrast, the locally run 'Residents' Association' deals with local issues including organising community functions, running the temple, dealing with utility companies and the administration of Indian ration cards, and holding the population records for the camp. As such, this Association is often described as equivalent to an Indian *panchayat*.

During the period of my research there was a relatively clear division of admin- istrative labour between the two authorities but the relationship between them was under negotiation. From the Residents' Association perspective, the Welfare Office was seen as the 'official' representatives for the colony, to whom I should speak first as this was the 'right order' and 'proper way' of going about things. The Welfare Office saw the Residents' Association as a bridge between themselves and the colony residents, but seemed wary of the authority that the Association held locally and how this impinged on its ability to control what happened in the colony. Thus, while TGiE policies and rhetoric have been actively encouraging 'self-reliance' and the devolution of authority to local levels, ironically, where this local leadership is already strong, such as in Majnuka Tilla, the TGiE appears to be uncomfortable and seeks to moderate it.

Besides Majnuka Tilla, several other settlements have also been established and managed independently of the TGiE. One such group of communities is the '13-Settlements' (*Tsogka Choksum*), a loose coalition of settlements established in the mid-1960s and populated mainly by refugees from Amdo, whose distrust of the Ü-Tsang-dominated TGiE stemmed from historically strained relations between central and eastern Tibet.[12] The largest of the independent communities was Clementown, established by the 13-Settlements leader Gontang Tsultrim in 1964 on land donated to him by Acharya Vinoba Bhave, the spiritual successor of Mahatma Gandhi. The strained relationship with the TGiE was expressed by an older Clementown resident when he explained that,

> in the early days the people asked for help from the Tibetan government but no help came… so people decided to do everything themselves so now we can continue to run the whole settlement and manage everything ourselves (April 2007).

The settlement continues to be run by the settlers with their 'Society', headed by Khochhen Rinpoche of Mindrolling monastery, managing a school and two clinics, dealing with the Indian authorities and receiving revenue from property rentals. Its administration therefore functions in a way similar to that of other settlements, but without a TGiE-imposed Local Assembly and Local Justice Commission. The settlement also has a different 'feel' from Dekyiling, the TGiE-run settlement located on the other side of Dehradun. Clementown is wealthier, with fewer restrictions regarding who can move into the settlement,[13] and settlers expressed a greater attachment to the settlement, perhaps because they chose to move there rather than, in the case of Dekyiling, being sent there by the TGiE.

Whilst there is no obvious TGiE presence within Clementown, the settlement does now come under the jurisdiction of the Uttarakhand CRO, and the secretary of the settlement's Society was keen to point out that many of the settlers now pay voluntary contributions to the TGiE (see Chapter 6). The 13-Settlements movement declined in the 1980s with a 'general tendency towards a new perception of national unity and the need for a centralised administration [and] an institutionalised and democratic system of government' (Ström 1995: 94). Most of the settlements are now reintegrated into the 'mainstream' of exiled refugee society under the authority of the TGiE, a trend that can be read, as several interviewees pointed out, as the TGiE and its institutions becoming progressively more powerful and spatially extended. These 'independent' settlements therefore illustrate both the inherent fragility of TGiE authority and the territorialising strategies it uses in its attempts to bring 'wayward' settlements under its control.

Rehearsing Tibetanness

What has been presented so far is an overview of the territorialising strategies that the TGiE employs to manage and govern these spaces. I now want to shift attention to the symbolic importance of the settlements as *national* spaces associated in complex ways with the homeland: an importance that is cross-cut with issues of temporality and rehearsal. In addition to the pragmatic aim of creating economically self-reliant Tibetan communities within the host state, the most frequently articulated rationale behind the establishment and maintenance of the settlements is that of protecting and preserving Tibetan culture, identity and way of life (CTRC 2003). 'Deliberately designed in such a way as to recreate Tibetan society with its core values intact' (Norbu 2001: 15), the settlements have been key to the exile government's project of reconstructing 'Tibet' in exile and, as the time in exile has extended, they have been framed as places where the community can pass on 'authentic' Tibetan traditions to the next generation. Echoing this official narrative, a number of interviewees spoke of the 'real Tibetan community' being in the settlements, which are in effect spaces where 'Tibetans can be Tibetans'.

As part of its nation-building exercise, in the first few years of exile the TGiE adopted a policy of locating refugees from each of the three regions of Tibet together in the same settlement (see Ström 1995). Mirroring the Nehruvian promotion of pan-Indianism and 'unity through diversity', this policy formed a key part of a wider promotion of a distinctly pan-Tibetan identity in exile that, it was hoped, would prevent communalism and promote national unity.[14] As Akhil Gupta and James Ferguson argue, place-making is ubiquitous in collective political mobilisation with '"homeland" ... remain[ing] one of the most powerful unifying symbols for mobile and displaced peoples' (1992: 11). However, the connection between the settlements and the Tibetan nation is not only imposed by the TGiE nor only discursive. Reflecting 'the complexity of... ways in which people construct, remember, and lay claim to particular places as "homelands" or "nations"' (Malkki 1992: 25), settlers themselves have also created the settlements as spaces of nationalism through both material and performative associations with the homeland.

Despite the majority of Tibetan settlements in India being located in physical landscapes starkly different from those of Tibet, each settlement is 'Tibetanised' through the recurrent use of Buddhist structures (e.g. Figure 4.2 and Figure 4.3). In addition to the cluster of Tibetan offices, schools and clinics that form the administrative core of each settlement, such 'scenery' and 'props' include monasteries, small temples for the protector deity of each village, archways decorated with auspicious Tibetan symbols, prayer-flags atop buildings and strung through trees, and *chortens* and *mani*-stones along roadsides. The consistency of these structures across different settlements is striking as, while the architectural style of the houses reflects the building materials and climate of their location within India, the recurrence of these cultural markers means that each settlement can be read in a similar way, and feels 'familiarly' Tibetan. As such, Majnuka Tilla was described to me by newcomer refugees as a reassuring Tibetan 'sanctuary' within the very Indian – and alien – city of Delhi. Thus, in reading the landscape of the settlements as socially constructed 'cultural texts through which political values are communicated and discourses enacted within particular societies'

Figure 4.2   Chortens, Lugsum-Samdupling. (Photo by author.)

(Till 2003: 349), the playing out of these symbolic links to the homeland can be seen as vindicating the TGiE's settlement programme, if simply in succeeding in the aim of recreating a 'mini Tibet' in exile.

The inscription of the Tibetan homeland onto settlement landscapes is not confined simply to the material environment, but is also evident in the everyday practices performed within these spaces. Illustrative of this is the traditional Tibetan practice of *kora*, a clockwise circumambulation of a temple or holy structure that is performed to venerate these sites and gain merit for the next life. *Kora* is central to Tibetans' use and understanding of space, with *kora* routes around *chortens*, temples and monasteries constituting ritual space within the settlements, especially for elderly residents (see Figure 4.3). With the Dalai Lama's temple and residence located there, Dharamsala has the most well-defined *kora* of the settlements in exile. As Christiaan Klieger notes, this 'capital' of Tibet in exile has a heavily scripted landscape as it is:

Figure 4.3   Large prayer-wheel, Dharamsala. (Photo by author.)

a place where memories and nostalgia for a lost way of life are perpetuated as no
other… The Dalai Lama's palace on the top of the hill, with Namgyal monastery
and the Tsuglhakhang nearby, with the people's village down the road, neatly reproduces the landscape of Lhasa itself… Such simulation is powerful and highly useful
as a mnemonic of the golden days of pre-1959 Tibet. (2002: 3; see also Anand 2002)

As a 'reprocessing of popular religious practices' (Fortier 1999: 50), the tradition
and practice of *kora* is a key 'place-making' strategy (Hewitt 1983) in the settlements and therefore adds a performative layer to the Tibetanisation of these
landscapes.

In sum, this network of settlements forms a core part of the political-
geographical imaginary of exiled Tibetans, 'conducive to the growth of memory

and the pursuit of the myth of return' (Colson 2003: 9). Thus, like the spaces of exiled Palestine and Western Sahara, Tibetan settlements in South Asia are 'vital political and symbolic spaces in the struggle for national liberation' (Farah 2009: 90; Ramadan 2009). Indeed, in one way these performances and enactments of Tibetanness reinforce the argument that nationalism – even when articulated in exile – requires some sort of territorial base in order to be articulated (Smith 1981). However, what is striking about this case is that there are other, unconventionally territorial, mechanisms through which the exile government has sought to reconcile the situation of displacement with maintaining strong *symbolic* links to the homeland of Tibet. Central to this is the parliamentary election system. Not simply a mechanism for conducting elections, the voting system for the TPiE was designed to play a key role in shaping the political consciousness of the diaspora and forging a symbolic link with the homeland. Moreover, the territorial organisation of TPiE elections encapsulates not only the strategies that the TGiE employs to connect multiple notions of territory within a regime of governance but is also a revealing site of experimentation where the administration is seeking to train its diaspora in democratic practices and rehearse electoral politics.

Whilst all parliamentary election systems are shaped by and, in turn, reflect the specific characteristics of the polity in question, there is a basic assumption that the national constituency will be divided into territorial units and that elected representatives will represent voters in their spatial constituency.[15] However, in the Tibetan case the exile parliament is organised according to a symbolic quota system where the dominant categories are regional and religious affiliations in the homeland of Tibet. The central element of this consociational schema is a regional quota whereby 30 parliamentarians are elected by voters in India, Nepal and Bhutan to represent each of the three traditional provinces of Tibet: Ü-Tsang, Kham and Amdo. Representation is split equally, with ten members elected from each region, and voters exercise their franchise according to the region in Tibet that they or their parents are from.[16] In addition, the four schools of Tibetan Buddhism – Gelug, Kagyu, Sakya and Nyingma – and the traditional Bön faith each have two elected members in the Parliament. Only monks and nuns registered with monasteries in South Asia can vote for candidates from their religious denomination, which effectively entitles them to a second franchise. Finally, to reflect the increasingly global Tibetan diaspora, four MPs are elected by Tibetans living in the West with voters in North America and Europe electing two representatives each. These four MPs are elected without reference to a traditional province or sect, making this the only conventionally territorial aspect of the TPiE system.[17] This electoral system thereby symbolically brings the territory of Tibet into the heart of TGiE politics, and offers an inventive way to overcome the limitation of organising political representation within a host state.

Underpinning the electoral system is the idea of symbolic territory. With regards to the broader electorate, even as Tibetans in Tibet are not enfranchised they are represented by proxy by the regional MPs. This symbolic representation of an otherwise disconnected population is thus a strategy through which the TGiE constructs its claim that it represents – and is legitimated by – Tibetans both in exile and in Tibet. However, the fact that the symbolic representation of Tibet, rather than the realities of life in exile, has formed the basis of the election system highlights a key tension at the heart of Tibetan democracy. As Jane Ardley argues, within the electoral system 'the notion of territory exists in the imagination and simultaneously in the past and the future – in "old" Tibet and "free" Tibet' (2003: 351). It is hardly surprising, therefore, that the operation of a symbolic electoral system in a diaspora of voters with local issues raises challenges and tensions. In interviews with members of the Tibetan electorate, criticism of this aspect of the system revolved around three interrelated narratives: the fact that representatives were accountable neither to 'real' constituencies nor to political parties with manifesto promises;[18] the perception that the system perpetuates divisive regionalism and creates identities that do not reflect the reality of 'modern' exiled Tibetans; and that such detachment leads to political apathy within the community (see McConnell 2013c). Underpinning the first of these critiques is the issue of accountability. As a teacher in Lugsum-Samdupling expressed it:

> If I vote for MPs then how are they accountable to me – answer me how? For they represent this half-real place of Kham – where I'm from – but what can they ever do in Kham? Sure it is occupied by the Chinese! But then, if I am in Bylakuppe and I have some problem with Indian police, then my MP cannot help me because their constituency is in Kham! (November 2007)

Another issue of contention is the fact that people from the three regions in Tibet and the five religious sects are generally scattered in exile, meaning that the de facto 'constituency' of any one elected representative is dispersed over settlements in India and Nepal. This lack of a fixed spatial constituency has, unsurprisingly, led to a significant communication gap between MPs and their constituents. Moreover, the issues faced by the majority of the Tibetan electorate in India often fail to be adequately represented within the TPiE due to the fact that most MPs reside in Dharamsala or Delhi rather than the exile settlements. Frustrations borne from such a detachment were expressed by a farmer in Sonamling who explained:

> Each Tibetan settlement in India has different problems and issues – for some it is high unemployment, for some drugs or alcohol abuse, conflict with Indians. Here our problems are with water shortage, infrastructure and [being] so remote ... but if most MPs are living in Dharamsala, then these issues they don't come in our parliament (May 2007).

Ironically, therefore, it is the mundane functionality of democracy on the ground in exile that is most limited in this case, with the symbolic territoriality of the electoral system creating significant challenges when faced with local territorialised politics.[19]

Remaining with issues of symbolic national territory, but returning to the settlements themselves, a further layer of complexity is worth noting. Not only have the settlements been constructed as islands of Tibetanness serving to preserve the 'original' homeland in the imagination, but also for second-generation Tibetans who have never seen Tibet and who have moved to Indian cities or to the West these Tibetan spaces in India in effect constitute a pseudo homeland. Such a displacement of ideas of 'homeland' requires a shift in conventional frames of reference: the role of the TGiE and its base in India has effectively created a 'domestic' population (those residing in Tibetan spaces in India and Nepal) and a (second) diaspora of Tibetans who have moved from South Asia to the West. As such, the TGiE acts as a 'home state', managing the transnational practices of their diaspora in the West, from remittances flowing back to India to participation in TPiE elections. Moreover, while the TGiE invests considerable material and ideological work in creating distinctions between an 'inside' and an 'outside', including the (re)creation of Departments of 'Home' and 'International Relations', the location of and relationship between the domestic and the foreign in this case is far from conventional. As such, this is a situation that confirms the assertion that exile politics disrupts the division between national and international politics, between inside and outside (Mandaville 1999).

This construction of the settlements as pseudo homelands is articulated and materialised in a number of ways. For example, there is a logical attachment to, and nostalgia for, the settlement where those born in India have grown up. This was expressed by a number of younger interviewees through support for sports teams in inter-settlement tournaments, alumni associations for settlement schools, and websites connecting those who have emigrated to the West with their 'home settlement'. More prosaic connections and attachments with exile settlements are also evident, with an increasing investment in property by some within the diaspora. A number of interviewees in Dekyiling, Clementown and Lugsum-Samdupling commented on how having a property in a settlement was seen as providing financial security, particularly if family members were living 'abroad'. Whilst the land within the settlements is managed exclusively by the TGiE and reverts back to the TGiE if or when the family leaves, there is an increasing trend of Tibetans expanding their homes and, in some cases, informally leasing additional land from local Indians.

Such emotional and material attachments to places and property in exile not only refutes assertions that diasporic relations to territory are confined merely to the imaginary and the nostalgic, but also highlights a crucial temporal issue within the community. What was intended as a temporary sojourn in exile is becoming increasingly permanent. This uprooted diasporic community is increasingly growing roots (see Fortier 1999; Malkki 1992). As such, this raises the important question of how easy and desirable it will be for the exiled community to 'up sticks' and leave India for Tibet should the situation in the homeland be resolved, or indeed if their position in India becomes untenable. One view, articulated by a young college graduate as we walked around the extensive grounds of the Tibetan Children's Village school in Dharamsala was that:

> Of course we will leave India if one day we have a free Tibet. When the Chinese invaded our people left everything behind and said that they'd be back in a few years – it's the same here. As we say, 'if it's written on your forehead then you go' – these are just buildings. People have a strong attachment to their homeland … even those like me who have never seen it. So of course we will leave here and go back (March 2006).

However, other respondents were sceptical that such an upheaval would be easy or feasible, arguing instead that the exiled community has become too established and comfortable in exile and has thus lost the desire to return to the homeland. Concern over and opposition to precisely such attachments to places and to lifestyles in exile has been a recurrent issue within the community. For example, an older resident in Dekyiling explained how:

> when land was first given to the Tibetans in South [India] the people they begged our leaders not to give them land. They thought you see … that having land in India, it would mean they would not return to Tibet, that they would always be in India (April 2007).

Likewise, a resident in Lugsum Samdupling recounted how:

> His Holiness came to Bylakuppe one time and saw how big some of the Tibetan houses were, so much bigger than neighbouring Indian houses, so he suggested to people that they don't spend all this money on houses, building new floors because it causes bad feelings with locals, and Tibetans are not in India to stay – one day we will have to leave everything here and go back to Tibet (November 2007).

Refusal to renovate dwellings 'lest this be misunderstood as acquiescence to resettlement' (Farah 2009: 89) is an act recurrent across exiled and refugee populations and, though many younger Tibetans differentiate between efforts to improve their living conditions and their commitment to the political struggle, fears of putting roots in exile continue to be articulated. Such opinions point to the fact that, for some within the exile community, the balance has tipped too far in the direction of providing for and settling the exile population to the detriment of furthering the struggle for the homeland. This was passionately expressed by a student in Majnuka Tilla:

> What is the purpose of our government? What are they doing for our cause? All they do is talk. We don't need these schools, these hospitals. All we need is [a] bed and food and to work for our cause. There is too much time wasting going on here. Our

people shouldn't be owning buildings, restaurants here – this isn't our country. One day India will no longer want us, and what then? (April 2007)

Whilst such heartfelt frustration with the direction that the community has been going in needs to be seen in the context of this individual's somewhat acrimonious relationship with the exile administration, it nevertheless foregrounds the conflicting temporalities that so characterise life in exile. The sentiments documented above can be read as attempts to *recover* a sense of liminality and transience. But speak to others within the community, and the experience of being in limbo is a very real and present one. From my time spent in the various Tibetan settlements, waiting as 'an active, conscious, materialized practice in which ... time and space often become the objects of reflection' (Jeffrey 2008: 957) was a recurrent issue, particularly among newcomer refugees and unemployed young people. Central to this, and echoing Chakrabarty's (1999) work on the inscription of waiting on the urban landscape, were particular spaces within the settlements such as tea stalls, *karom* boards and bed-sits where groups of young Tibetans would regularly 'hang out'. These individuals spoke of such places as simply for 'timepass'. They were places to put in the hours, days, months, or even years before they got their chance to return to Tibet, or emigrate to the West. Dharamsala in particular is dominated by these often male-dominated spaces of timepass, which have in many ways shaped a distinctive 'Dhasa' subculture.

So what we have here, essentially, is a tension common across refugee settlements between idleness, contingency and enduring temporariness (Agier 2002; Ramadan 2012) and the perception that emotionally and materially investing in these spaces brings the danger of losing the hunger for – and sacrificing the right to – return to the homeland. But I want to suggest that the TGiE's material and symbolic construction of the settlements as spaces in which the practices of governance can be rehearsed and where state personnel can be trained offers a way of beginning to ameliorate these temporal challenges. Writing about Gaza, Eyal Weizman argues that 'an improved camp with ... functioning institutions, updated physical and communication infrastructure and better homes, is not a negation of the right of return but rather an instrument for its reinforcement' (2011: 146–147). In the Tibetan case the TGiE's emulation of state practices of territorialisation, bureaucratic administration and welfare provision adds another dimension to this reconfiguration of refugee settlements as political spaces, where political mobilisation can not only be articulated, but where material and performative preparations for futures elsewhere can be rehearsed.

Conclusion: Territorialising Exile Governance

I have sought to outline in this chapter how, though decidedly not a conventional case of centralised power operating over absolute, bounded territory, the TGiE does achieve more than simply exercising diffused power over deterritorialised, transnational networks. This exile administration's construction of territory and power is somewhere between 'state space' and 'diaspora space', articulating 'deterritorialisation and reterritorialisation at the same time in the same spaces' (Ó Tuathail & Luke 1994: 382). In summarising the arguments made here I want to return to Joe Painter's work on territory as a state effect and suggest three ways that this case speaks back to and extends his argument. First, given the TGiE's position as a 'temporary' guest in the host state of India, territory as the '*outcome* of networked socio-technical practices' (Painter 2010: 1090, emphasis added) has particular characteristics in this case. Here, territory as an effect is unstable, highly contingent and co-produced by both Indian and exile Tibetan 'state' actors. More precisely, it is contingent on Indo-Tibetan relations (which are, in turn, dependent on Sino-Indian relations), on the political innovations devised by TGiE to function as a government despite these territorial insecurities, and on the interactions between Indian and TGiE officials at a range of scales. The latter are underpinned by relational networks around land tenure, policing of settlement boundaries and settling intercommunal disputes, all of which vary considerably according to the local politics at the Indian state level. Indeed, as I discuss further in Chapter 7, understanding the geographies of this (non)state formation offers a novel insight into the complexly differentiated nature of Indian politics, from the latter years of the Nehruvian era to contemporary Hindutva dominated politics.

Secondly, territory as a state effect in this case is both material and symbolic. What the TGiE lacks in material territory in India (and has been denied in the homeland in Tibet), it seeks to compensate for by incorporating symbolic territorial associations into its governmental practices. This is evident both in terms of the symbolic role of the settlements as 'mini-Tibets' and in the TGiE's parliamentary election system. The latter is not only a revealing site of experimentation where the administration is seeking to rehearse electoral politics, but also a mechanism through which TGiE seeks (and often struggles) to connect material and symbolic notions of territory. The complex temporalities and spatialities of Tibetan democracy links to a third way in which this case speaks to questions of territory. As Doreen Massey (2005) notes, we live in so-called 'spatial times', and with the rehearsal spaces of the settlements having an acute sense of 'place-temporality' (Wunderlich 2010), this case offers an opportunity to rethink the relationship between time, territory and the state. This is certainly not to collapse time and space into each other, but rather to delineate the practices and narratives of what is an unusual spatio-temporal configuration of governance.

Central to this are the resonances that the TGiE's (re)territorialising strategies have with state power and state space (Brenner et al. 2003; Jessop et al. 2008). As sketched out in this chapter, the TGiE employs a number of state-like territorialising mechanisms to circumvent and ameliorate the challenges raised by governing dispersed spaces. These include the centralisation of authority in Dharamsala, attempts to ensure uniformity of governance and facilities across the settlements, and the creation of multi-tiered jurisdictions across which state functions are performed. Through these territorialising strategies the TGiE is striving to institutionalise the Tibetan nation in exile at a range of scales. However – and there are always 'howevers' in this case – the TGiE's lack of official sovereignty over these spaces means this is far from *legal* jurisdiction. There is a territoriality to TGiE's governance practices, but this is de facto rather than de jure control over territory. Given these legal limitations and internal challenges to the administration's governmental reach it is important not to overplay the case for TGiE's 'state space'. Moreover, in supporting the assertion that territory is neither the sole preserve of nor 'inherently tied to the state' (Elden 2005: 8), the aim of TGiE's territorialising strategies are also based on the specific exilic objective of fostering a sense of Tibetan nationalism and unity of purpose across the diaspora in South Asia. Carrying with it notions of a deliberate strategy of 'creating' territory in the exiled population's imagination, 'strategic territorialisation' (Boudreau 2001) is perhaps a useful descriptor for how TGiE manipulates notions of territory for tactical political goals.

The settlements can thus be seen as inherently ambiguous spaces; as condensation points for the contradictions of exile, where the debates and realities of deterritorialisation and reterritorialisation, of temporality and permanence are played out and become visible. They are spaces where the focus on preserving the past and planning for the future means that the present effectively gets squeezed. Thus, although the situation of any exiled population is exemplified by the dilemma between meeting immediate needs by putting down roots in exile, and retaining the emotional and political desire to return home by maintaining a sense of restlessness, in the Tibetan case the discourses and practices through which these debates are played out are intricately linked to broader issues of territory and governance.

Endnotes

1   Tibetan Buddhist structure around which circumambulations are made.
2   See Shri S.A. Dange's comments during the 'Discussion re: situation in Tibet, *Lok Sabha*, 8 May 1959' (TPPRC 2006: 64).
3   A number of settlements are located in what has since become Uttarakhand, established in 2000. Mainpat settlement lies within what is now Chhattisgarh state.
4   Population figures for the settlements are sourced from interviews at the Department of Home (September 2012).
5   Around 4000 Tibetans were granted asylum and settled in Bhutan in the 1960s. However, following an alleged attempted coup by Tibetans in March 1974, relations between the refugees and the Bhutanese authorities deteriorated (Norbu 1976). The Bhutanese Royal Government subsequently demanded that all Tibetans in the state take Bhutanese citizenship and assimilate into the host community. Whilst several hundred accepted this citizenship, many Tibetans believed losing their refugee status

would contradict their ultimate goal of returning home, and were rehabilitated in Tibetan settlements in India in 1980.

6   Dharamsala is in effect two towns: Lower Dharamsala is a predominantly Indian market town while Upper Dharamsala, or McLeod Ganj, is where the majority of the Tibetan community is based. The headquarters of TGiE is located midway between the upper and lower towns at 'Gangchen Kyishong'.

7   Whilst there are similarities in terms of the dispersal of refugees, the Indian government's relatively rapid handover of governance of the Tibetan settlements to the TGiE sets this case apart from the disastrous efforts to rehabilitate East Bengal Hindu refugees in West Bengal in the early 1950s (Chatterji 2007).

8   As the Indian Constitution grants land property rights only to citizens and the majority of Tibetans in India do not hold Indian citizenship, these individuals are barred from privately owning land. However, some Tibetans in Himachal Pradesh have circumvented this limitation through what is known as a *benami* transaction, where land is illegally purchased in the name of a silent Indian partner, often from ethnically Tibetan tribal regions of Spiti or Lahaul. In a move to regularise these land transactions and as 'a special welfare measure keeping in view the Government of India's policy guidelines regarding resettlement/rehabilitation of Tibetan refugees' (Order No.Rev.B.F.(10)199/2003, 8th May 2006), the Himachal government issued 50-year land leases to TGiE in 2005, which the exile government can then sub-lease to institutions or Tibetan individuals (District Commissioner, Kangra, November 2007).

9   A religious challenge to the leadership of the Dalai Lama and the TGiE is also posed by a small minority within the exile community who follow the protective deity Dorje Shugden (see Mills 2003b).

10   For example, Drepung and Ganden monasteries near Lhasa have been rebuilt in Mundgod settlement, and Tashi Lhunpo in Shigatse and Sera outside Lhasa have been re-established at Bylakuppe.

11   The name 'Samyeling' is used primarily in official Tibetan contexts. To residents, Tibetans who pass through it and Indian 'visitors' this area is known by its local name of Majnuka Tilla, often shortened to 'MT'.

12   This suspicion ran both ways with the TGiE critical of the 13-Settlements' independent operations and acceptance of funds from the Commission for Mongol and Tibetan Affairs in Taiwan, with tensions intensifying during the 1970s (Ström 1995). Although the 13-Settlements coalition was disbanded and relations are much improved, this remains a sensitive issue within the exiled community.

13   Several families and individuals from Himalayan regions such as Ladakh, Sikkim and Lahaul also reside in Clementown.

14   However, even within TGiE-run settlements there is often (self)segregation of regional and sectarian groups. This is manifest in an unofficial division of communities into 'villages' based on where the settlers are from in Tibet: a micro-geography that is distinct from TGiE designated administrative blocks.

15   Notable exceptions are elections to the Israeli Knesset where the entire country constitutes a single electoral constituency.

16   Where parents are from different regions, the child chooses which of these two regions to elect from. Candidates are put forward by regional associations, which act as

pseudo-constituencies although, if elected, the supported candidates have no formal obligation to their regional association.

17 According to the 1991 Charter Tibetans living in Australia, Japan and Taiwan do not have a representative and so have no franchise in these elections. More generally, proportional representation has been rejected in favour of a focus on equal representation through this quota system.

18 Based on the argument that political parties would cause problematic divisions within the diaspora the exile Tibetan case is a rare example of non-party democracy with parliamentarians functioning as both the incumbent and the opposition.

19 An alternative electoral system based on an upper house with members from regions and religious sects, and a lower house with elected members representing constituencies in exile and based on proportional representation, has been discussed but there has been little serious consideration of it in recent years.

Chapter Five
Playwright and Cast: Crafting Legitimacy in Exile

Leaving the Tibetan settlements in India briefly, I want to take you to a university seminar room in central London on a hot July evening in 2011. The room was booked by a Tibetan student after requests on social media for an emergency community meeting and was packed with members of the UK Tibetan community – old and young, monk and lay, those with British citizenship and those recently arrived from India. The topic for debate and discussion: the legitimacy of the exile Tibetan government. Three months earlier the Dalai Lama had announced his retirement from politics, and the week before, the TGiE had been dissolved and reconstituted as the 'Institution of Tibetan People'. This was a move that, to many of those attending, was an act of their government voluntarily de-legitimising itself. The discussion was lengthy, animated and impassioned, and raised the following questions: Where does legitimacy come from? Can it be lost, retained, regained? Who has the right to say that a governing institution is legitimate? How can you gauge whether people deem an authority legitimate when their freedom of expression is denied?

This chapter seeks to set these questions in context, and to use the broader framing of legitimacy – who claims it, and how it is (re)constructed on a daily basis – to examine from another angle the modes of authority articulated by the TGiE, and the extent to which these are framed as state-like. With the previous chapter examining the extent to which this polity enacts forms of centralised and territorialised authority, here attention shifts to questions of leadership, instrumental authority and the construction of rational-legal authority (Weber 1978). These articulations of state-like infrastructural power (Mann 1984) are viewed

Rehearsing the State: The Political Practices of the Tibetan Government-in-Exile, First Edition. Fiona McConnell.
© 2016 John Wiley & Sons, Ltd. Published 2016 by John Wiley & Sons, Ltd.

through the lens of the various roles adopted and prescribed within this rehearsal state. Thus, with the previous chapter having 'set the scene' for this rehearsal state, giving a feel for the spaces of this polity and some of the administrative structures it has developed, this and the following chapter are structured around three takes at *peopling* this 'state' in exile (Jones 2007): the Dalai Lama as playwright; the training and professionalisation of exile government bureaucrats; and, in Chapter 6, the various roles assigned to – and relationships forged with – the exile population.

'Legitimacy' is a term that dominates official and public discourses on international politics. It has framed discussions of regimes overthrown during the Arab Spring (Lynch 2012), international interventions in Bosnia, Afghanistan and Libya (Jeffrey 2013; Pack 2013; Rubin 2007) and is central to the institutions of global governance (Erman & Uhlin 2010). In academic circles, scholars from a range of fields have examined the contested concept of political legitimacy. In brief, international law views legitimacy in an institutional recognitional sense in terms of whether or not a polity is recognised by other sovereign states as having met the international legal criteria for legitimate statehood (Montevideo Convention 1933; Crawford 2006), whereas political philosophy generally understands legitimacy in a normative sense as the right to govern and thus a status conferred by the people on the government (Sternberger 1968). Meanwhile in political science issues of legitimacy have been linked to those of consent, with scholars such as Lipset (1984) and Easton (1975) arguing that legitimacy involves the capacity of the political system to engender the *belief* that existing institutions are the most appropriate ones for the society. With the exile Tibetan case being one where international recognition – and thus political legality – has been denied, this polity will never meet the criteria for legitimacy under international law readings. However, as I set out in this chapter, the significant claims for political legitimacy made by the exile Tibetan leadership can be taken seriously if we turn to both normative arguments around legitimacy in principle and empirical questions of legitimacy in practice (Wilson & McConnell forthcoming).

The most celebrated typology of legitimacy is Max Weber's (1978) tripartite formula of charismatic, traditional and rational-legal authority, with legitimacy based on the personality of a leader, on popular customs and on constitutional principles respectively. This schema continues to dominate political science approaches to legitimacy and, based on the assertion that charismatic and traditional sources of authority are increasingly redundant, there has been a pervasive state-centrism to most accounts (Barker 1990; Jessop 1997; Migdal 1988). Put simply, this is the argument that the state plays a key role in securing political legitimacy and that, in turn, legitimacy is central to the appraisal of the modern state. However, accounts based on historical transformations of ideal types of legitimate authority often fail to interrogate what actually constitutes legitimacy and how it is fostered, negotiated and maintained (Alexander 2011). In light of this and the more tenuous claims this exile polity has to statehood, my focus here is on the *process* of legitimation itself. This opens up two productive lines of enquiry. First, it means that we can take a 'broader look at the applicability of [legitimacy] to other groups besides formal governing institutions' (Horowitz 2009: 249) – to, for example, the degree of legitimacy of legally unrecognised polities (Wilson & McConnell forthcoming). Or to put it another way, what has been framed as a problematic ambiguity within the concept of legitimacy – that it is almost impossible to establish a clear-cut distinction between a 'legitimate' and an 'illegitimate' regime (Shain 1989) – is actually a productive ambiguity. Second, in shifting away from seeing legitimacy as something that a state 'secures' to a process that is always under contestation this brings to the fore the importance of everyday practices of statecraft.

In what follows I am interested in statecraft as a form of knowledge of production – particularly in the field of bureaucracy – and will primarily focus on the people who craft the state: both the leaders and the state personnel. Treading a careful line between using Weber's typology as a starting point for raising pertinent questions and not taking it as a prescriptive framework, this chapter begins by tracing two sources of the TGiE's legitimacy as articulated through dominant narratives: the claim that the TGiE is not a new institution established in exile, but the transplantation of the Ganden Phodrang founded in 1642; and the role of the Dalai Lama as a quintessentially charismatic leader. This section explores how the Dalai Lama has in effect taken on the role of playwright in this rehearsal of statecraft in exile by instigating transformations of the exile government institutions and practices. Particular attention is paid to democratisation in exile, an arena wherein the exile polity is rehearsing and in many ways improvising modes of governance. The second section of this chapter focuses on the role of bureaucrats in reproducing this rehearsal state and the increasing importance of rational-legal modes of authority both to this exile polity's claims to legitimacy and to the honing of statecraft. Finally, attention is paid to the development of expertise, the training of these bureaucrats and the implications that recent professionalisation initiatives have for the remit and future of this rehearsal state.

The Dalai Lama as Playwright: Charismatic Leader and Democratic Visionary

Between the seventeenth century and 1959 the Dalai Lamas – a lineage of religious leaders of the Gelug school of Tibetan Buddhism – were both the religious and political leaders of Tibet and headed the Lhasa-based Tibetan government (see Chapter 3). As the personification of *Chenrezig*, the bodhisattva of wisdom and compassion and the protector deity of Tibet, the role of the Dalai Lama in defining Tibetanness, embodying Tibetan culture, and providing continuity to the history of Tibet cannot be overemphasised (Kolås 1996). Moreover, given the lack of centralised governance in pre-1959 Tibet, the personal qualities of the Dalai Lama assumed heightened importance, and this continues today, with His Holiness functioning 'as the central locus of power and identity within the Tibetan diaspora' (Houston & Wright 2003: 218). Born in Amdo in 1935 and recognised as the Dalai Lama when he was two years old, the leadership role of the 14th and current Dalai Lama has, due to the political turmoil in Tibet in recent decades, proved to be of heightened significance compared to his predecessors. This remains the case in exile where, denied access to the homeland, Tibetan refugees place immeasurable importance on the physical and moral presence of the Dalai Lama.

But what of the nature of the Dalai Lama's *authority* more specifically? Carole McGranahan argues that the immense respect and devotion shown by Tibetans to the 14th Dalai Lama indicates that 'this is not a case of an authoritarian state or a situation of "dominance without hegemony", but instead one in which authority, in the form of the Dalai Lama, is deeply hegemonic' (2010: 4). In focusing specifically on the Dalai Lama's authority vis-à-vis the TGiE I want to tease out two aspects of this hegemony: His Holiness's role in legitimising the exile project, and his visionary leadership in scripting exile politics. First, as the medium through which political authority and national identity are enacted and secured into the future, the Dalai Lama is a key source of Tibetan legitimacy. More precisely, it was frequently repeated to me that it is the legitimacy of the exile government – its institutions, practices and ability to represent the Tibetan people – that is embodied in the person and the institution of the Dalai Lama. A senior TGiE official perhaps best summed this up when he stated: 'the true legitimacy we [TGiE] have – where our sovereignty is located – is in the moral authority enjoyed by the Dalai Lama' (September 2010). But through what mechanisms does the Dalai Lama's authority bestow legitimacy on the exiled administration? For a start, this is a form of legitimacy that is not, and arguably has never been, defined in territorial terms or through the direct use of force. Rather it is a discursive mode of power enacted over a defined population (Wilson & McConnell forthcoming). And the source of this authority is, as Anne-Sophie Bentz argues, 'the symbolic importance attached to the *institution* of the Dalai Lama' (2012: 294, emphasis added). Redolent of the sovereignty of monarchs in medieval Europe (cf. Kantorowicz 1957/1981), the Dalai Lama is simultaneously a person and an embodiment of *Chenrezig*. Both the personality of Tenzin Gyatso, the 14th Dalai Lama, and the cycle of reincarnation, is thus key to the power of influence that His Holiness holds. It is the concept of charisma that goes some way to capturing this dual nature of the Dalai Lama's authority.

Max Weber defines charisma as a social status rather than as personal qualities: 'a certain quality of an individual personality by virtue of which he [*sic*] is considered extraordinary and treated as endowed with supernatural, superhuman, or at least specifically exceptional powers or qualities' (1978: 241). As the incarnation of *Bodhisattva Avalokiteśvara* the Dalai Lama inspires immense loyalty and compliance from Tibetans and is thus arguably an archetype of charismatic authority. However, there is an in-built vulnerability to such authority for two reasons. First, following Weber, this authority is a social status that is conferred on an individual by the belief of the people: recognition of charismatic authority can be withheld as well as granted. Second, charisma is reliant on a single individual. As the Dalai Lama himself stated in a speech to the Tibetan Parliament-in-Exile:

> No system of governance can ensure stability and progress if it depends solely on one person without the support and participation of the people in the political process. One man rule is both anachronistic and undesirable (14 March 2011).[1]

In order to seek to mitigate these vulnerabilities at the heart of Tibetan leadership, and mindful of his own progressing years and strained relations with Beijing, the Dalai Lama has implemented a series of changes designed to downplay the political authority derived from his personhood and the institution that he embodies. Central to this is the Dalai Lama's role as the key instigator of the quintessential foundation of rational-legal authority: democracy. His Holiness summed up his commitment to democracy in a speech to the British All Party Parliamentary Group on Tibet in March 1991 where he stated:

> Change is also coming to the Tibetan political system. It is unfortunate that it happens in exile, but this does not stop us learning the art of democracy. I have long looked forward to the time when we could devise a political system, suited both to our traditions and the demands of the modern world ... I believe that future generations of Tibetans will consider these changes among the most important achievement of our experience in exile. (TPPRC 2003: 2)

With the Dalai Lama using his 'extraordinary personal qualities ... to generate enthusiasm for particular state projects and strategies' (Jones 2007: 23), it is this promotion of democracy that has arguably been the most significant role His Holiness has played in terms of the cultivation of stateness in exile. As such, the Dalai Lama's role is redolent of that of a playwright. I suggest this analogy rather than directorship because not only has the Dalai Lama sought deliberately to step away from a more directorial role in exile politics (discussed below), but playwright points both to a visionary mind-set and to craftsmanship in skilfully conceiving and orchestrating socio-political relations.

Hailed as the greatest achievement of the exile community, the democratisation of Tibetan politics is central to how this exile administration presents itself to the international community. As discussed in Chapter 7, democratisation can be read as the Dalai Lama and his administration aspiring to emulate Western ideals of 'good governance' in order to prove their credibility and legitimacy to the international community. However, democracy is also a key arena in which the TGiE rehearses and improvises modes of governance. In Tibet prior to 1959 the spheres of governance and politics lacked mechanisms for political participation (see Chapter 3). After assuming the temporal and spiritual leadership of the Tibetan state in 1950, and before his flight into exile, the 14th Dalai Lama sought to introduce a number of democratic initiatives through his Reform Committee,[2] but his initiatives were obstructed by the Chinese occupying forces. Therefore, whilst there are antecedents to democracy in pre-1959 Tibet, Tibetans had no direct experience of democratic governance when they came to India.

The relative political freedoms of exile, and the Dalai Lama's interactions with Indian and Western legislators, helped His Holiness refine and implement his vision of democracy based on the union of participatory politics and Buddhist values. From the establishment of the first exile parliament in September 1960 until 1990 was a period of gradual democratic reforms. The first parliament saw the introduction of principles of democratic governance such as the separation of powers among the three branches of legislature, executive and judiciary. However, political changes during this period were predominantly symbolic, had limited impact on Tibetan society, and the Dalai Lama remained the ultimate arbiter in political decision-making. The 1980s saw a number of abortive attempts to develop democracy in exile, political scandals in Dharamsala and the rise of a number of organisations run along regional and factional lines that were, according to Tibetan author and political commentator Jamyang Norbu, 'basically reactionary, and their influence on society unhealthy and divisive' (2004: 19). In light of such challenges, the Dalai Lama introduced a number of major reforms in 1990 in order to accelerate the process of democratisation. After dissolving the exile parliament and suspending it for a year, His Holiness reinstated it as a fully fledged legislative body with an expanded membership, independent authority and effective powers over the executive. In a move that Dhundup Gyalpo described as '"a great leap forward" that ushered in a new and more mature phase in the governance of the Tibetan administration' (2004a: 26), the *Kalon Tripa* (now *Sikyong*) has been directly elected by the Tibetan diaspora since 2001.[3]

Apparent from this chronology is the fact that Tibetan democracy is significantly different in evolution and form compared to most Western democracies. Though it is usually prudent to avoid assumptions of Tibetan exceptionalism (Anand 2006), in the case of exile Tibetan democracy descriptors relating to 'uniqueness' are genuinely apt. The most unusual aspect of Tibetan democracy is the fact that rather than the conventional process of democracy being driven by popular demand through uprisings and armed struggles, or imposed by a foreign power, in the Tibetan case it was instituted from the top down by the Dalai Lama. Indeed, Tibetan democracy is often referred to as a blessing or gift from the Dalai Lama, with a primary school teacher I spoke to in Sonamling, Ladakh, describing how His Holiness 'gave democracy like Buddha giving teachings'. As Maria Edin (1992) points out, traditional literature on democracy simply fails to take into account a situation where a leader wishes to give away his power due to

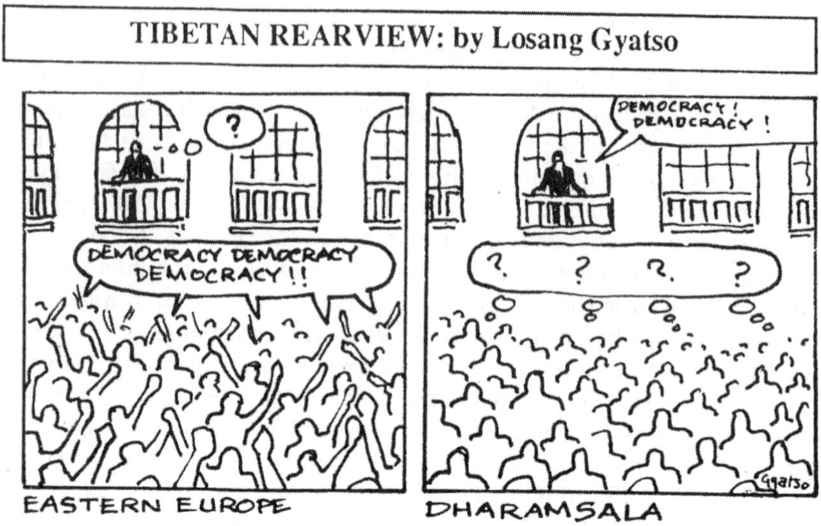

Figure 4   Cartoon from the *Tibetan Review*, January 1992, p. 11.

his convictions. This reversal of the norms of democratisation is deftly illustrated by Losang Gyatso's cartoon, published in *Tibetan Review* in 1992 (Figure 4).

At first glance, this case seems to be an ideal model of democratic transition, as it was instituted both peacefully and by the existing leadership. However, ironically, the very fact that Tibetans in exile did not experience democratisation through first-hand personal struggle means that the democratisation process has been far from smooth or straightforward. As a number of politically active interviewees asserted, a gifted democracy is impossible to refuse, but without asking for it in the first place it is also difficult to appreciate and appropriate: it is 'like a solution in search of problems'. The benign imposition of democracy by the Dalai Lama onto a society that was arguably not yet ready for such a socio-political transformation has therefore led to a weak form of democracy. This has been manifested in a lack of public awareness of the core principles of democracy, a reluctance to assume decision-making and leadership responsibilities, and low participation in democratic practices. The institutions of democratic governance are thus in place, but there has, until the 2011 elections, been a general reluctance to rehearse the practices of democracy.[4]

A second key stumbling block to the realisation of a 'fuller' democracy, even defined in the exile community's own terms, is precisely the unique patron-deity status of the Dalai Lama. Being an unelected monk, the Dalai Lama's political legitimacy is based on his 'special wisdom' and the obedience he enjoys from his citizens. This clash of embodied charismatic authority, and rational-legal democratic governance sits uncomfortably with liberal democratic ideals and appears anachronistic to a Western audience (Ardley 2003). It is a situation that also means that many Tibetans participate in democratic activities primarily because the Dalai Lama asked them to do so, and not because they want influence over the government or to alter the polity. Indeed, the Dalai Lama's bestowment of democracy has been far from universally welcomed by the exile community, with His Holiness facing resistance from many to what they saw as the diminished role of their revered leader. Such attitudes, and the influence of Buddhism on exile Tibetan politics more generally, have, however, been challenged by some amongst the younger generation. Though careful not to directly challenge the Dalai Lama's authority, a number of young Tibetans I spoke to echoed the sentiment of a Tibetan graduate in Delhi who described how 'for Tibetans, whatever the Dalai Lama says, they will believe and go along with, for them his word ... is Tibetan law ... this is a mental and political block in our community and to progress our politics we have to change this' (September 2011).

In many ways these challenges to the maturing of exile Tibetan democracy affirm more general assertions that democratisation is a protracted and fragile process, which is both difficult to teach and requires a process of gradual evolution in order to take root (Dahl 2000; Held 2006). However, what is significant here are the important strategic and temporal reasons behind the Dalai Lama cajoling the diaspora into engaging with democracy over the past decades. Crucially, the transition to democracy has been framed by the Dalai Lama and, more recently, the *Sikyong*, as both offering security for the present and hope for the future. Underpinning both are concerns about the continuity of political legitimacy.

Starting with the more distant future, in its exile manifestation the Dalai Lama's rolling out of democratic structures is a calculated strategy to create an active democracy-in-waiting. As outlined in the TGiE's second Integrated Development Plan:

> A democratically elected, representative government, with an open, accountable and efficient administration, will be one of the greatest needs of future independent Tibet. Professional expertise in conducting elections will be invaluable in devising appropriate modes of democratic governance after Chinese occupation has ended, and in undertaking the necessary reconstruction of the nation. (Planning Council 1994: Section 9.3.9)

In portraying the development of democracy as offering hope for a future democratic Tibet in the longer term, the exiled leadership at this time thus framed democratisation in exile as a chance for the community to experience and practise democratic governance in anticipation of implementing such a system within a future Tibet (shifts in this vision for future Tibet are noted in Chapter 7).

Addressing both longer-term concerns around the continuity of legitimate authority and short-term concerns around political stability was the landmark decision made by the Dalai Lama in March 2011 to retire from political life and transfer his political authority to elected leaders. In standing down as both head of state and head of government – though retaining his role as spiritual leader – the Dalai Lama dissolved the historical form of Tibetan government (the Ganden Phodrang) and thus relinquished an almost 400-year-old tradition of power.[5] This devolution of authority to the TGiE has been framed as a key stabilising mechanism in light of an imminent power vacuum after the current Dalai Lama passes away. His Holiness describes the decades of democratic reforms in exile as paving the way for this transition, in particular striving to inculcate a sense of individual political responsibility amongst exiled Tibetans and reducing their dependency on the Dalai Lama. It is also a transition that, the Dalai Lama asserts, 'will help sustain our exile administration and make it more progressive and robust' (19 March 2011).[6]

The Dalai Lama's retirement also raises questions of where legitimacy lies and how it is constituted, as well as the relationship between secular 'modernity' and religious 'tradition' (see McConnell 2013b). Given the Dalai Lama's role as a key source of Tibetan legitimacy, not only have Tibetans in the diaspora and in Tibet been reluctant to support His Holiness's resignation from politics, but his decision – and the amendments to the Charter that have formalised it – pose important constitutional questions for the TGiE. Indeed, it created something of a 'legitimacy crisis' within the exile community. When the Dalai Lama sought refuge in India in 1959, he declared: 'Wherever I am, accompanied by my government, the Tibetan people recognize us as the Government of Tibet' (cited in Dulaney et al. 1998: 41). In the intervening years, the TGiE has claimed a clear legal continuity with the pre-1959 Tibetan government, a continuity that has not been threatened by the democratic reforms outlined above. However, in severing the political role of the lineage of the Dalai Lamas, which underpinned the TGiE's traditional authority, and renaming the TGiE in Tibetan as *Tsenjol Bod Mei Zhung Gi Drik Tsuk* ('Institution of Tibetan People'), it appeared that the exile administration was relinquishing this claim to continuity. The question thus arose as to whether this was in effect a retraction of the TGiE's legitimate authority, and the transformation of the TGiE into a body representative only of Tibetans in exile. It was precisely such issues that were vexing the Tibetans' meeting in London that I mentioned at the start of this chapter. While the line from the exile leadership is that it is business as usual, and that the name change is an administrative amendment to appease disgruntled Indian bureaucrats rather than a substantive change in the nature of the administration, lingering concerns nevertheless remain. These were the focus of a lively discussion I had with Tibetan students at Jawaharlal Nehru University (JNU) in Delhi. One particular student who had grown up in Tibet summed up the issues thus:

the idea of legitimacy is so important for our community now and for this we are seeing different forms of legitimacy being discussed in the open. With the Dalai Lama's retirement and name change of the government, the question being asked is

PLAYWRIGHT AND CAST: CRAFTING LEGITIMACY IN EXILE

'has TGiE lost all of its traditional and charismatic legitimacy? Or does it still retain some of these?' I think that some traditional legitimacy has been retained despite the change of official status. Tibetans still believe that TGiE is the continuation of the pre-1959 Tibetan Government, and most outsiders, they are oblivious to the name change because it is in Tibetan! With charismatic leadership, there are two sides to this. On one side there is the problem of attaching so much to a single leader – cult of the individual, and vulnerability when that person is no longer around. On other side, the international community look for a single leader for a cause like ours. This is where the Tibetan diaspora is in a strong position. I think strong leadership is important, but it should be focused on the *role* of the *Kalon Tripa*, not the individual, not attached to a personality, and so more sustainable (September 2012).

As I write, it is still early days for this transition of political authority from the Dalai Lama to the position of the *Sikyong*, but some emerging concerns are worth noting. On the one hand, it is arguable that neither the intertwining of the religious and the political, nor the political role of the Dalai Lama, has been fully rejected in this recent shift of power. Article 1 of the amended Charter states that His Holiness remains the 'guardian and protector' with the powers to 'provide advice and encouragement' to the Tibetan people, to 'provide guidance' to the government, and to be an international spokesperson 'on behalf of the Tibetan people'. Indeed, the Dalai Lama is still looked to for guidance on political issues, from the self-immolations in Tibet to the machinations of exile politics.[7] However, on the other hand, the Dalai Lama's retirement means that the current *Sikyong* is expected to take on much of the political authority previously borne by His Holiness.

Elected in 2011, Lobsang Sangay, a Harvard law scholar who was born and educated in India, therefore plays a significantly more prominent role in the Tibetan movement than his predecessors. With such a background, Sangay represents an important shift in exile leadership and brings with him a new style of Tibetan politics: one that is young, Western-educated and secular'.[8] Sangay's more overt political ambition for power has not only injected a new energy into exile Tibetan politics, but it also hints at the emergence of a more directorial role for the *Sikyong*. As hinted at by the JNU student cited above, there also appears to be an attempt to transfer charismatic authority to the role of the *Sikyong*, albeit certainly not to the extent held by the Dalai Lama. In Weber's understanding of charisma, the archetypal form of 'genuine charisma' – the 'authority of the extraordinary and personal *gift of grace*' (Weber 1919/2009: 79, emphasis in original) can be transformed into 'charisma of office', an attenuated form 'that would have to be combined with interest' (Bentz 2012: 303; see also Shils 1965). Such a transfer of charismatic authority to the institution of the *Sikyong* is, however, contingent on the rational-legal authority of the exile administration, and it is to the increasingly important role of this form of state-like authority that I now turn.

Rational-Legal Authority and the Construction of a Bureaucracy in Exile

Of the rational-legal coupling, it is the latter part that the TGiE struggles to live up to. As discussed in the previous chapter, though a Supreme Justice Commission has been established and regional and local justice commissions are in the pipeline, these institutions are a symbolic element of the TGiE's three pillars of democratic government rather than having any substantive legal authority (Duska 2008). Given the TGiE's lack of recognition and its functioning within the state of India, the Justice Commission has the legal standing of an arbitration body operating under the Indian Arbitration and Conciliation Act of 1996. It can deal only with civil disputes within the community and, even then, without legitimate powers of physical force or detention the punishments it can dole out are limited. These include 'victim compensation', social service, formal apologies, providing offerings to particular deities and, in more serious circumstances, the withdrawal of an individual's Tibetan citizenship and the voting rights and access to TGiE welfare and services that this permits.

In a similar vein the 1991 Charter of the Tibetans-in-Exile cannot be legally binding on individual Tibetans and is not in and of itself a legal document, not least because it needs the approval of the majority of the Tibetan (i.e. those from the homeland as well as diaspora). However, as with constitutions of conventional states, the Charter does offer insights into both the political culture of the exile Tibetan polity and the formal rules by which it is organised. Promulgated by the main 'playwright' in this case – the Dalai Lama – the Charter outlines the main institutions of the 'state' (the executive, legislature and judiciary) and enshrines the rights and obligations that formalise the relationship between political authority and the people. Modelled on similar documents from liberal democracies and complying with the laws of India, the Charter combines international norms of good governance with principles of Tibetan Buddhism. However, given the situation of exile, this document also encapsulates some of the temporal and legal challenges that this community faces. It is deliberately called a Charter not a Constitution not only to avoid raising eyebrows in New Delhi but also to reflect that this is a document intended for the interim period of exile. It is certainly spoken about as a model for how a future democratic Tibet might function and as part of a broader programme of training Tibetans for future governance, but it is not a blueprint and it would be annulled if or when the diaspora returned 'home' and replaced by a 'proper constitution' for the territory of Tibet.

Yet there is, of course, a limit to how much a constitution can tell us about a state, conventional or otherwise. Such documents are statements of intent written by political elites, and say little about the everyday practices that constitute governmental authority. In order to explore the political culture of this community in more detail I want to turn to a system at the core of Weber's idea of rational-legal authority (1947: 153–154), and the mechanism through which formal rationality in governance is delivered: the bureaucracy. My starting point for focusing on bureaucracy comes not from debates as to whether it is characterised by 'the social production of indifference' (Herzfeld 1992) or whether a recovery of the ethics of the bureau and bureaucrat is needed (du Gay 2000). Rather, my interest stems from the perspectives on the workings and the effects of the state and on political legitimacy that it offers. As Colin Hoag argues, attending to the 'beliefs, practices, and limits of state functionaries' can 'provide an important counter-weight to recent, prominent interpretations of the coherent, "seeing" state (Scott 1998), or of the "magical," "illegible" state (Das & Poole 2004; Taussig 1997)' (Hoag 2011: 87). The lens of bureaucracy thus offers a key insight into the contingency of state authority and the everyday construction of legitimacy; moreover, the training of Tibetan bureaucrats in the crafts of the state is, I want to suggest, central to the project of rehearsal.

The topic of bureaucracy has traditionally been 'the domain of sociologists and political scientists' (Hoag 2011: 81), but there has been increasing interest in bureaucratic processes from geographers and anthropologists (e.g. Beer 2008; Heyman 1995; Shore & Wright 1997). As Merje Kuus notes, this focus correlates with an interest in 'the "how" of politics – in the daily practices rather than the formal scripts of political struggles' (2011: 423). Underpinning such research is an application of ethnographic approaches to explore the everyday workings, challenges and dilemmas of bureaucracies and bureaucrats (Ferguson 1990; Herzfeld 1992; Strathern 2000) and an examination of the materiality of bureaucratic worlds (Gordillo 2006; Riles 2006). Drawing on such approaches to explore the exile Tibetan bureaucracy, I also return to Weber and, in particular, the series of features that he asserts defines a public bureaucracy: the conducting of administration through defined hierarchies and according to fixed procedures; the vocational aspect of public service; everyday habits of civil servants that constitute 'reiterative authority' (Feldman 2008); organisational culture; and the development of bureaucratic expertise (Kuus 2014; see also Wilson & McConnell forthcoming).

Hierarchies, demarcated responsibilities and fixed procedures are at the core of modern bureaucracies, and the exiled leadership has made concerted efforts to integrate these elements within the TGiE. As such, the TGiE itself is appraised by the 'Office of the Auditor General' and the exile civil service has defined roles, training schemes and pay-scales standardised across the diaspora. Reflecting the idea that 'diffuse forms of power often seek rhythmic conformity and spatio-temporal consistency through the maintenance of normative rules and conventions' (Edensor 2010: 11), working hours and holidays of government offices are consistent across the dispersed settlements and scattered communities. In terms of staffing, until 1972 the recruitment and appointment of Tibetan civil servants was overseen by the 'Service Management Office', a division of the Home and Security Department. With the expansion of the TGiE a separate office known as the Department of Personnel was established under the supervision of the *Kashag*, and this, in turn, was replaced in the early 1990s by the Public Service Commission. With regards to the responsibilities of this body, the Public Service Commissioner explained:

> Our main job is the recruitment, promotion and training of staff, and disciplining or dismissing staff if there has been improper discharge of duties. In the Commission we have responsibility for officially appointed staff. We have around 3000 which is less than a single Indian Government office! This includes locally recruited staff in sanctioned posts, but excludes teachers and doctors and local staff appointed in the settlements by the Settlement Office. For these people the official appointee, the Settlement Officer, he [*sic*] looks after the staff welfare, pensions and promotions (September 2012).

Recruitment procedures vary according to the type and rank of the post, with higher level positions being filled primarily through promotion, and direct recruitment being used to fill junior positions. Jobs are advertised in the TGiE's Tibetan and English language publications, Tibetan newspapers, on Tibet TV and online. Meanwhile staff promotions are considered twice a year and are based on conditions set out in the 'CTA Staff Statutory'. Alongside regularised promotion procedures, pay scales are also standardised across the TGiE. A series of pay commissions were established from the mid-1990s with the assistance of senior Indian officials. As one of those involved explained, before the first commission:

> everyone seemed to be paid something different. 'Tashi' would be earning 9715 [Indian] rupees per month and in the next office, doing similar work, 'Tenzin' would be earning 7355 rupees – and there would be no explanation for this! (October 2012)

The pay commission interviewed staff from across the government, and 13 ranks of posts and pay scales were established. Employment pass books and service books were then issued to all civil servants so that hours and pay could be logged. At the micro-level, the 'Audit Manual' for staff specifies in detail the 'Allowances and Compensation' for exile government staff, from the class of train ticket that different ranked officials are permitted to claim in travel expenses, to the per diem paid to TGiE staff during official visits – from 225 Indian rupees (Rs) per day for staff at Joint Secretary level and above in 'non-remote' areas of India, to Rs. 160 per day for peons and drivers. As a senior TGiE official involved with these reforms noted:

> This was the first time in Tibet's history that we had a civil service set out in this way with structured roles, pay grades and career paths. With these changes we had a lot more transparency for the government. It meant we looked like a government! (September 2011)

As explored in the previous chapter, these definitive features of bureaucracy – of 'fixed rules and procedures, within a clearly-established hierarchy and in line with clearly demarcated official responsibilities' (Pierson 1996: 20) – underpin how the TGiE attempts to centralise and extend its authority over the dispersed settlements. The fact that most Settlement Officers are posted on three-year stints has drawbacks in terms of these individuals sometimes lacking 'local' knowledge and not being in post long enough to see major projects to completion. However, their being indisputably 'government staff' and their rotation between settlements means that there is significant consistency of practice across these spaces. A cadre of trained administrators who have developed expertise in their roles has thus been fostered over the decades. Facilitating this are regular reports to be submitted to Dharamsala, annual Settlement Officer meetings where changes to TGiE policies are discussed, and a 'rule book' issued by the Department of Home outlining in detail their role and responsibilities.[9] Standardised government practices are also ensured across other departments. For example the Secretary at the Department of Health described how:

We have written guidelines for all hospitals and rules and regulations for all our staff – doctors, nurses, pharmacists, administration. Also all pay scales are the same and are organised from the Department of Health and the utility of hospital vehicles, it all comes under the same umbrella (March 2006).

Government staff in the settlements are thus fully familiar with the systems and philosophy of the central government and are able to implement relatively standardised management practices.

Back at the TGiE headquarters in Dharamsala government posts are designated as either contract or permanent, and post transfers are assigned annually, with notification given to individuals through coloured documents – yellow for the former and pink for the latter. As a mid-level official in the Department of Information and International Relations explained:

this day when they give out the slips can be very nervous for people – say if they are working in one department here and their wife is teaching at upper TCV [Tibetan Children's Village School in Dharamsala] and suddenly they are transferred and have to go serve in South [India –so it is such a move, and they have no choice in this if they want to stay in the government. So I suppose ... this kind of thing means it is acting in a very government-like way with such staff transfers. Other organisations wouldn't make these demands (September 2011).

Tibetan officials I spoke to in a range of posts and grades, described job transfers as a defining feature of their working and personal lives. After decades of service, it is not uncommon for senior staff to have held posts in almost all the TGiE departments and worked in several TGiE offices overseas. Meanwhile mid-level officials are often rotated around the settlements. For example, one official who was a Settlement Officer when I interviewed him talked through his 32 years of service, starting as a land surveyor in the settlements in Karnataka, then as a cooperative manager in Tezo, Arunachal Pradesh, and a handicraft centre manager in Darjeeling before being posted to Settlement Officer at Dekyiling. As he put it, he had been a 'Tibetan government tourist seeing most of India!' This is a system that, seeking to emulate the perceived norm of states rotating their civil servants around embassy postings and local government departments, means that high-level cadres know the community, the locations where Tibetans are living, and the TGiE itself in significant detail. However, while this practice of rotation proves productive at the level of settlements, its implementation across the network of Offices of Tibet has been critiqued in light of the community's limited human resources (see Chapter 7). Put simply, the TGiE has neither the quantity nor quality of staff trained in diplomacy in order for such a rotation system not to be detrimental to its foreign policy goals. As a former TGiE official expressed in frustration:

> we can't afford to lose skilled, experienced people, but the government insist on sticking to the procedures which are copied from the Indian and other governments. It makes no sense in our case. We can't afford to have someone completely new come to important posts like in Brussels and Geneva and then have to learn the UN or the EU system from scratch and build up networks and build up trust – this takes months and years for someone to learn. This urgent time for Tibet is not the time for Tibetans to learn, it is the time to act (March 2012).

The system of transfers also takes its toll on family life. Narrating a similar story to that of the Settlement Officer mentioned above, a businessman now based in Nepal recounted how his family followed his TGiE official father from Mainpat settlement in Chhattisgarh to Dalhousie, then to Mundgod settlement in Karnataka and back to Dalhousie, then to Orissa, Mussoorie, Kathmandu and finally Pokhara. This peripatetic lifestyle was 'really hard for all our family, with moving house, leaving friends, changing schools. Each of my siblings, they were born in a different place! But there's no choice – when the government tells you to move, you move' (June 2007).

Such personal sacrifices bring to the fore the role of vocation. The framing of office-holding within a civil service as a vocation is central to Weber's analysis of bureaucracy, with state personnel perceived to be acting not in a personal capacity but rather subject to a particular sense of public duty (Weber 1978: 220–221). In the Tibetan case this was evident in terms of the Public Service Commissioner's description of a 'good government staff' being 'hard working, and putting maximum of what they know into the role. They must think first of the government and the community, and second about self' (September 2012). Echoing Yael Navaro-Yashin's (2006) analysis of the desire Turkish-Cypriots have for government jobs in their unrecognised republic – despite their critique of the 'state' – jobs in the TGiE are highly sought after, not primarily for their salaries (though posts do come with accommodation, medical cover and pensions), but as a meaningful and valued way of contributing to the exiled community and to the wider Tibetan cause. TGiE staff frequently spoke of pride in 'serving the government and their people', and were at pains to mention how many years of service they had already given. One young official in the Dharamsala Welfare Office described his as a 'job for life', which, with the significant responsibilities that came with it, 'should be taken seriously' and meant that he had chosen a different lifestyle from his friends and peers who had not joined the civil service. Indeed, a number of officials in Tibetan schools, hospitals and cooperatives spoke of their dedication to their job as the material manifestation of their commitment to the wider Tibetan cause: it 'gives me inner peace of happiness to be working in our community and contributing to our struggle' (Secretary, Sonamling Settlement, May 2007).

However, the role of an increasingly sophisticated bureaucracy in a national freedom struggle is in many ways a double-edged sword. On the one hand, quotidian bureaucratic practices can have a significant stabilising effect in what are often turbulent times in exile. According to Ilana Feldman this was certainly the case in Gaza during the regime changes under the British Mandate (1917–1948) and Egyptian Administration (1948–1967). On the other hand, an argument I often heard expressed by those who are politically active but not TGiE employees was that the increasingly 'bureaucratic nature' – meant in a pejorative sense – of the exile government resulted in time, energy and resources that could have been invested in the freedom movement being 'wasted' keeping the TGiE itself running. Echoing the plea noted above to act not learn at this 'urgent time for Tibet', there was frustration amongst some in the community that Tibetan civil servants were stuck in the minutiae of bureaucratic routines in ways that meant that the longer time-frame of return was fading from view. The pragmatics of the rehearsal were diverting attention from aspirations of a 'final performance'. As a Delhi-based Tibetan journalist put it:

> Our government is such a bureaucracy these days – it's so hard to get anything done quickly. Everything has to go through so many different committees and levels of bureaucracy. Everywhere there are protocols, for every small project in any settlement it has to go through the Planning Commission ... and then approval is needed from the *Kashag*. OK so there is transparency, but it means the departments, they have no time or people to look at the bigger scale issues, at the policy level (September 2012).

Despite such frustrations, Tibetan civil servants wield a not insignificant degree of authority within the exile community. Describing how he deals with Tibetans, foreign tourists and local Indians on a daily basis, the Dharamsala Welfare Officer explained:

we have to do the right thing every time ... you see we are the representatives of the government, and if we do something wrong then it will look really bad for all the Tibetans. We are the public face (April 2007).

On one level, it is this status as 'representative' of the TGiE and, by extension, the Dalai Lama, that endows these officials with the moral (rather than legal) authority to mediate disputes and make key decisions within the settlement communities. This is a form of authority that thus draws on the hierarchies and paternalism of the traditional political system (Goldstein 1975), and is manifest in significant ways. Perhaps the most obvious example is the substantial powers held by Settlement Officers, who have the authority to issue official TGiE letters and documents, provide official TGiE stamps and signatures and can thereby recommend (or oppose) an individual for scholarships or visa applications. However, acceptance of the authority of Tibetan bureaucrats is not always unconditional, and generational differences are apparent. As a young man in casual employment in Majnuka Tilla recounted:

the older generation, they see the Welfare Officer and other government staff as representatives of His Holiness – like the staff from Lhasa. So no matter if that person is a devil or a saint, still they will bow to him and stick out their tongue.[10] They are always showing so much respect, even if they don't agree with what he says. But then ... the younger generation, we are more straight. If I don't agree with the Welfare Officer, I will tell him directly (June 2007).

Yet, despite being increasingly willing to challenge government officials when judging them to be competent, younger Tibetans nevertheless largely accept their claim to authority. This belief that the Tibetan public has in the authority of their government's bureaucrats thus indicates that another form of power is also at work here: 'it is a strength of bureaucracy that it can produce its own author-ity' (Feldman 2008: 90). In documenting the production of governing practices in early twentieth-century Gaza, Ilana Feldman traces this authority to a series of everyday practices that she describes as 'repertoires of authority' (2008: 93; see also Hansen 2005). Speaking to Richard Sennett's notion of authority as a 'process of interpreting power' (1980: 20), this repertoire includes the repetitive actions of issuing, processing and filing documents, alongside the development of a particular style of rule through the everyday habits of civil servants (Feldman 2008: 16).

Elements of such 'reiterative authority' is certainly in evidence in the exile Tibetan bureaucracy. TGiE offices across India share in common a number of administrative and aesthetic features, from the rows of labelled files on the walls to panoramas of Lhasa, photographs of the Dalai Lama and maps of India on the walls, low chairs covered with small Tibetan carpets, and letter headed paper and stamps with the government's seal. There is also striking standardisation across the appearance and comportment of TGiE staff (cf. Goffman 1959). Female officials wear the traditional Tibetan chuba while their male counterparts are either in the male equivalent or a white shirt with dark trousers, and small Tibetan flag pin badges are often in evidence at official functions. While enacting a distinctly *Tibetan* identity, these civil servants also 'look the part' in ways that speak to images of modern, technocratic governance, in stark contrast to the elaborate ceremonial costumes worn by Tibetan government officials in pre-1959 Tibet. There is thus a banal stateness to these habits, aesthetics of government offices and styles of dress, and this is underpinned by a distinct 'organisational culture' (Goldman 2006; Weber 1947), which unifies the government workforce through the standardisation of linguistic codes and bureaucratic communication (see Bourdieu 1994).

The state-like qualities of Tibetan bureaucracy can, in part, be attributed to its emulation of the host's modern, state bureaucracy. Indian bureaucrats were instrumental in advising the exiled leadership in the early days and continue to provide training for Tibetan officials. The ranks within the Tibetan bureaucracy match closely the hierarchies in the Indian system, as do pension provision and the 'Annual Confidential Report' for each state employee. The Public Service Commissioner traced the roots of TGiE bureaucratic structures back even further, noting that the exile government was 'practicing the British-Indian system' and was in effect the 'proud grandchild of the British system!' However, officials were keen to differentiate between their emulation of the Indian *system* of bureaucracy and their eschewal of their hosts' bureaucratic *practices*. Whilst many Tibetans I spoke to noted the creep of 'Indian bad habits' around the increasing layers of officialdom and the length of time to process identity forms, the TGiE certainly has an impressively clean slate in terms of corruption, with stringent transparency and accountability systems in place. A story circulating in Dharamsala in summer 2011 was that it took one department so long to get various invoices signed off for a new hard drive that by the time permission was granted, the hard drive had been hacked!

A central aspect of organisational cultures is the conformity of individual staff to institutional codes and mores. In the case of the TGiE this has, in recent years, been expressed most explicitly through the leadership's insistence on civil servants' support for the Middle Way Approach for the future of Tibet (*Umaylam*; see Chapter 7). Attempts have been made, through written and verbal warnings and the threat of dismissal, to bring into line government employees who express dissatisfaction with the government line and/or support *rangzen* (independence). Consequently, some TGiE officials spoke of 'wearing two hats', one their role within the TGiE, the other their own opinion on the future of the homeland, and of having to be increasingly careful in keeping the latter separate from their professional life. Such tensions are a microcosm of the broader challenge of seeking to represent a freedom struggle through the mode of democratic governance and a state-like bureaucracy. However, the situation is not one of paralysing stasis:

improvisation can be an important element of rehearsal and it is to the adaptive and dynamic nature of exile Tibetan bureaucracy that I now turn.

Professionalising the Exile Tibetan Bureaucracy

Questions of expertise offer a useful spotlight on how this bureaucracy has evolved in over five decades in exile. As a number of scholars have argued, specialised knowledge and expertise are key foundations for the authority vested in civil servants and a cornerstone of rational-legal authority (Pierson 1996; Weber 1922/2009). Writing about the administrative revolution within Whitehall between 1830 and 1914 that created the 'technique' of Victorian government, Roy MacLeod argues that it was the advent of the expert – 'a protean image of authority and rational knowledge' (1988: 1) – that defined the birth of modern statecraft. Aspirations precisely towards 'modern statecraft' are at the heart of the story here, but what is of particular interest is the journey in reaching this goal. The pursuit of expertise, training of bureaucrats and professionalisation of the workforce are useful lenses through which to view important socio-institutional transitions within the TGiE. These transitions are partly explained by the maturing of the TGiE as an institution but are also shaped by the temporalities and spatialities of exile. With the increasing reach of the exile government, both across the dispersed settlements in South Asia but also increasingly stretching to the diaspora in the West, there has been, to echo Timothy Mitchell's work on governance in twentieth-century Egypt, a 'reorganization and concentration' (2002: 36) of knowledge within the community. Exile has provided the time and access to resources to acquire new skills, solicit advice and rehearse bureaucratic practices.

Before discussing the development of bureaucratic expertise in more detail an important question to address is who is actually making up the state personnel in the TGiE. Those who were appointed to positions in the various offices and welfare institutions that constituted the nascent administration in the early 1960s reflected closely the first generation of exile elites. These were mainly monastic leaders and Ganden Phodrang officials who had fled Tibet with or just after the Dalai Lama, and members of the former Tibetan aristocracy who had moved to northern India earlier in the 1950s (Kolås 1996). The latter had the advantage of being educated in India so, with a working knowledge of English and Hindi, were invaluable in establishing relief operations for the thousands of Tibetan refugees arriving in the subcontinent in the 1960s. For many years the higher echelons of the TGiE – the *Kalons*, Secretaries, Additional Secretaries and *Chitues* – were thus not only from the generation who had been born in Tibet, but were also largely either clergy or from established, noble families (including the Dalai Lama's family). For example, as there was virtually no campaigning by parliamentary candidates until the 2011 elections, *Chitues* were elected largely on the basis of who they were, the contacts they had and, in some cases, what family they were from.

This continuity of traditional political structures and patronage relations was not always apparent to external audiences, but critiques of the dominance of familial dynasties within the upper ranks of the administration and a degree of nepotism have been a relatively regular feature in the exile press (e.g. in *Tibetan Review* and *bod-kyi-dus-bab* ['Tibet Times', a Tibetan language newspaper published in Dharamsala]).

As the years in exile have lengthened, predictable shifts in the make-up of the bureaucracy have occurred. Underpinning these have been the rapidly improving educational standards within the community and the increasing embrace of that stalwart of rational-legal authority: meritocracy.[11] This is reflected in changes in TGiE recruitment, with admission to the civil service now dependent on tests in Tibetan and English, numeracy, computing and management skills, and a lengthy interview process. Mirroring the Indian context, the qualification levels required to be considered for TGiE jobs have also increased significantly in recent years. As a Tibetan career counsellor in Delhi explained: 'for a government job you now need at least a BA and probably an MA, and for pion jobs it used to be minimum 8th grade, now it's minimum 12th grade and basic English' (September 2011).

In the early days of exile, when few in the community had such qualifications, the Tibetan leadership was heavily reliant on Indian and Western experts for projects as diverse as establishing and running schools, clinics and cooperative societies, training Tibetan civil servants, conducting democratic elections and formulating integrated development plans (Planning Council 1994: Section 2.6.4.2). A recurrent theme when talking with senior or retired TGiE officials was the often complete lack of training that they had received. Reflecting somewhat lightheartedly on his own experience, the Public Service Commissioner explained how:

> Many years ago I was asked to take a job in one of the accounts units but I have no background in accountancy! Yet the personnel department they have looked and they think that among all the staff I have the best skills so I am appointed there for some time. It was the same with Health – after some time in the US I was posted to the Department of Health, but I have no background in medical at all. Yet they said, 'Ah, but you have spent time in the US and you know how they manage their health system so you can say how we can improve our system!' (September 2012)

This lack of training was a trend echoed in the judiciary where, for many years, most of the senior staff did not have law qualifications, but were 'simply civil servants posted to this division within our government' (Supreme Justice Secretary, March 2006).

However, changes have been afoot. Whilst the TGiE continues to rely on external experts in a range of fields, the prolonged length of time in exile and a drive towards self-sufficiency have seen the leadership increasingly seeking sources of expertise from within the diaspora. Justice Commissioners must now have a law qualification or be a practising advocate conversant in the Indian legal system, and Lobsang Sangay's administration in particular has focused on building human resource capacity. The *Sikyong's* nomination of *Kalons* in 2011 was especially revealing. On one level it was a box-ticking exercise in terms of appointing individuals from different regions, generations and genders (Tibetan Political Review 2011). However, attention focused on the appointment of Dikyi Chhoyang as DIIR *Kalon* and Tsering Wangchuk as Health *Kalon*. The former came from a community development post in Montreal and has experience of working in Tibet and Beijing, and the latter – a doctor trained in Poland – was the first medical practitioner to hold the post.

Yet a significant challenge facing this promotion of meritocracy within the TGiE is staff retention. As an official at the Planning Commission put it:

> it's hard to get people who are trained and will stay. The turning point for our community here in India was 1991 when the first batch went to the US. At that time we saw it as a good idea, that this was an opportunity for us. But we didn't calculate the huge impact it would have on our institutions – the impact both socially and publicly – and we are still dealing with this. It has decimated our institutions here in India. We have an attrition rate of 50%. You go to any department and they will tell you how many senior, experienced staff they have lost to the West. We spend all this time training staff and then after one or two years they go. How can we have sustainability [in a situation] like this? (October 2012)

One strategy to encourage staff retention has been to better match an individual's skills and interests with departmental posts, which has been implemented through an enhanced 'Recruitment Training Academy' run by the Public Service Commission. In this scheme graduates are selected to first receive five months of training in TGiE rules and regulations, accountancy, Tibetan history and international relations before being invited to apply for government posts in the department of their choice. Whether this attempt at enhancing capacity building within the TGiE is successful remains to be seen, but it certainly foregrounds a key concern of the administration: that of promoting professionalisation.

Two contexts underpin the recent prioritisation of professionalisation within the exile administration. First, high graduate unemployment levels in India (Jeffrey et al. 2008) and restrictions faced by Tibetans seeking to enter the Indian job market have led to increasing numbers of young Tibetans migrating to the West in search of employment opportunities. This is an issue of considerable concern to the exile leadership who, as discussed in the previous chapter, see the tightly-knit communities within the settlements in India as key to the preservation of cultural and national identity. The development of strategies for job creation and retention both in the administration and within the communities in South Asia is therefore a high priority for the TGiE. Also underlying this promotion of professionalisation are the recent shifts in Tibetan democracy. The Dalai Lama's decision to devolve his political powers is viewed by many as a call for Tibetans to assume more administrative and leadership responsibilities. Reinforcing this, in his inauguration address Lobsang Sangay explicitly called for the professionalisation of the Tibetan administration to 'ensure greater access and transparency' and stated that the TGiE would 'strive to reach 10,000 professionals among 150,000 in exile and appeal to Tibetans inside Tibet to reach 100,000 in the next two decades' (8 August 2011).[12] While Lobsang Sangay was not the first exile Tibetan leader to promote education and professionalisation – the Dalai Lama prioritised education from the early days of exile – nevertheless what is new here is both the focus on 'professionalisation', and strategies to link the diaspora in the West back to the community in India.

Emblematic of this is an initiative launched in 2012. 'Tibet Corps' links Tibetan professionals based in Europe and North America – from lawyers and accountants to engineers, epidemiologists and surgeons – with pro bono service opportunities in the TGiE (www.tibetcorps.org). Promoted as a 'reverse Peace Corps', suggested projects range from setting up a medical insurance scheme to developing renewable energy projects in the settlements and establishing a postal ballot system. Though at the time of writing this initiative is in the early stage of development, it raises a series of questions around diaspora mobilities, forms of governmental expertise and knowledge transfer. Firstly, premised on the transfer of social capital from the diaspora in the West to the communities in India the project is a strategy to influence mobilities and – to an extent – the geographical imaginations of exile Tibetans, seeking to strengthen ties of identity and solidarity across the exile Tibetan community. Though the internet has been a significant factor in connecting this dispersed population, the emphasis in this initiative is on Tibetans giving back to the exile community in practical ways through face-to-face contact and on-the-ground work. Secondly, underpinning the rationale of Tibet Corps is the aim of strengthening the TGiE and transforming the exile government's personnel from amateur administrators to state(like) bureaucrats. Promoting the development of social capital, the initiative thus has the potential to play a role in further training Tibetan bureaucrats in arts of the state and the practices of governance.

Questions arise, however, as to how this initiative will intersect with existing training of TGiE officials. How will Tibet Corps' unpaid volunteers from the West fit alongside the rigidity of existing TGiE bureaucratic structures with their defined roles, hierarchies and pay scales? Related to this are issues around competing models of expertise and which forms of knowledge are valued over others. With young, Western-educated professionals being endorsed as agents of social change, what space is left for other voices? Where, for example, do Tibetans from Tibet and their expertise in Chinese language and the situation in the homeland fit into this picture? The notable lack of such individuals holding positions within the TGiE was a recurrent issue raised by both 'newcomer' refugees and political activists. This situation is slowly beginning to change, with more Tibetans from Tibet being employed in junior positions and slowly working their way up the TGiE ranks, but questions remain as to *who* the TGiE is representing and serving, and for what purpose?

The relative stability of the exiled Tibetan community in India to date has certainly facilitated the training and professionalisation of TGiE bureaucrats and intellectuals of statecraft. In many ways it adds weights to the claim that exile and the waiting that it entails can be productive – providing valuable time to experiment, train, seek advice and develop a Tibetan political philosophy. However, what does it mean to have been in training and practising these crafts of the state for over 50 years? We now have a second generation in training, but for what end? The promised future back in the homeland seems to be increasingly slipping from view and the prospect of TGiE bureaucrats actually enacting these roles 'for real' in Tibet is no longer really on the cards. But if these officials are enacting their roles for real for the exile community, are we seeing perhaps the administration becoming 'a permanent government-of-exiles' (French 1991: 200)?

Conclusion: Constructing Legitimacy Through Crafting a 'State'

I want to conclude this chapter by addressing the question of *why* the exile Tibetan administration would want to emulate an often derided red-tape-bound style of state-like bureaucracy, with its rigid hierarchies, set working patterns and fostering of vocation? The answer, I have suggested, lies in the fostering of legitimacy. This is legitimacy not originating from political legality in the form of recognition, but rather constructed through the TGiE's claims to represent the Tibetan people. These claims are rooted in its traditional authority as the continuation of the Government of Tibet, the charismatic authority of the Dalai Lama and the belief and consent that it engenders from its population. With the exile government unable to leverage economic capital or physical force, it is the symbolic capital of bureaucratic rationalism and democratic governance that is prioritised in this case (Bourdieu 1994). Through this emphasis on cultivating and promoting rational(-legal) authority and democratic credentials, legitimacy in this case emerges as a mode of statecraft rather than an achieved status (Wilson & McConnell forthcoming).

Focusing on the processual nature of legitimacy also highlights its relational qualities: legitimacy is as much about subjects' belief in the validity of a political authority as it is about that authority's claims to be a rightful power (Barker 1990). Engendering of belief and consent vis-à-vis the state does not rely solely on looking or acting the part. Rather, legitimacy also requires buy-in from the population over which a state claims authority. As such, the following chapter shifts attention to the construction of political subjectivity through this exile polity, and focuses on the third 'peopling' of this rehearsal state: that of the role of the Tibetan population itself.

Endnotes

1   For the full text see: http://www.dalailama.com/messages/retirement/message-to-14th-assembly
2   This was a committee of 50 monks and lay officials who had the task of suggesting and implementing progressive changes to the administrative set-up in Tibet (see Gyalpo 2004a).
3   The 2001 *Kalon Tripa* election was won by Professor Samdhong Rimpoche with 84% of the votes polled. He was re-elected in 2006 with almost 91% of the vote.
4   The 2011 parliamentary and *Kalon Tripa* elections had the highest participation levels to date, with 49,184 Tibetans (59% of the exile electorate) casting votes in the final round, and lively hustings held in Dharamsala and on Tibetan radio and online.
5   Though the political responsibilities of the Ganden Phodrang have been removed, it remains as the institution of the Dalai Lamas.
6   Remarks made during a public teaching at the main temple (*Tsulagkhang*), in Dharamsala: http://www.dalailama.com/messages/retirement/retirement-remarks
7   It is notable that comments that the Dalai Lama makes publicly about the TGiE tend to be rapidly adopted by the exile parliament, a prime example being the shift in title from '*Kalon Tripa*' to '*Sikyong*' in 2012. See: http://tibet.net/2012/09/20/parliament-amends-kalon-tripa-to-sikyong/
8   The term 'secular' (translated as *remey*) is a somewhat confusing one in the Tibetan context as it is defined not as the absence of religion but as the state not discriminating among different religions (Shiromany 1998: 272).
9   The 'Guidelines for Settlement Officers' booklet was first issued by the Department of Home in 2003 and was to be found in each of the Settlement Offices I visited.
10 Sticking out one's tongue is a Tibetan gesture of respect and supplication (see Dresser 1997).
11 Illustrative of this shift towards meritocracy is the current *Sikyong*, Lobsang Sangay, who, in his election campaign, presented himself as an 'ordinary' Tibetan who could be any settlement resident's son.
12 For the full text see: http://tibetoffice.eu/statements/inaugural-speech-of-kalon-tripa-dr-lobsang-sangay/

Chapter Six
Scripting the State: Constructing a Population, Welfare State and Citizenship in Exile

In seeking to explore the nature of the TGiE's stateness, the previous two chapters have focused on particular perspectives vis-à-vis the sources and construction of state authority: the relationship between territory and power, and development of rational(-legal) authority, respectively. This chapter offers a third take on the question of *how* the TGiE seeks to rehearse state-like governance by focusing on the enabling forms of power that make individuals into subjects. This is therefore a shift from examining the instrumental value of rehearsal – exemplified in the previous chapter through the training of exile bureaucrats in the arts of statecraft – to attending to the constitutive effects of rehearsing the state in terms of creating Tibetan political subjects. Given that debates about the nature of political subjectivity are vast and often divergent (e.g. Arendt 1958; Bayart 2007; Edkins et al. 1999) what follows is the navigation of a particular state-focused route through this topic. I have the space here neither to trace a genealogy of Tibetan subjectivity per se nor to focus on the multifarious and sometimes divergent subject positions that are held simultaneously by individuals within the diaspora (see Anand 2000; Norbu 1992; Yeh 2002). Nor, indeed, is this an exercise in tracing out Partha Chatterjee's notion of the 'politics of the governed' vis-à-vis this case. Rather, I am specifically interested in the 'social rules, sanctions and prohibitions' (Pile & Thrift 1995: 2) through which political subjects are constructed by the TGiE and, in turn, how the governance of these subjects constitutes a form of stateness.

Denoting the micro-political practices through which a governing agency conditions people to act in specific ways and through which people govern themselves (Foucault 1991), the concept of governmentality offers a useful lens for how the

Rehearsing the State: The Political Practices of the Tibetan Government-in-Exile, First Edition. Fiona McConnell.
© 2016 John Wiley & Sons, Ltd. Published 2016 by John Wiley & Sons, Ltd.

individual is connected to the state and vice versa. What is particularly expedient in bringing the concept of governmentality into dialogue with the case of TGiE is the ambiguity surrounding the extent to which it is a state or non-state practice. On the one hand, an important element of Foucault's analysis of governmentality is the idea of the 'governmentalization of the state' (1991: 103) – 'the process by which the juridical and administrative apparatus of the state come to incorporate the disparate arenas of rule concerned with the government of the population' (Dean 1999: 2). In light of this, the majority of literature that draws on the idea of governmentality has viewed it as a technique of *state* power, focusing on how the state is formed through everyday practices of governance at different scales and what kinds of political subjectivities these governance projects seek to produce (e.g. Barry et al. 1996; Corbridge et al. 2005; Hansen & Stepputat 2001). On the other hand, governmentality is also an 'expansive way of thinking about governing and rule in relation to the exercise of modern power' (Watts 2003: 13). As such, it both posits power as process that continually needs to be enacted, and refuses the reduction of political power to the actions of the state (Miller & Rose 1990). This therefore facilitates a focus on 'the diverse and heterogeneous agencies' through which governance works (Dean & Henman 2004: 483), with governmentality understood as a multidimensional and trans-scalar endeavour that can be undertaken by a range of non-state as well as state actors (Legg 2005; Sidhu & Christie 2007). In sum, the concept of governmentality facilitates a focus precisely on the politico-legal ambiguity of the exile Tibetan polity.

Crucially, not only does governmentality open up the possibility of polities other than states governing but also it allows for entities other than territory to be what is governed. Whilst governmentality is an inherently spatial concept constituted of a range of territorialising strategies, Foucault's writings on governmentality in many ways underplay the role of territory (Elden 2007). Foucault asks whether the emergence of governmentality means that there is 'a shift of accent and the appearance of new objectives, and hence of new problems and new techniques ... from a "territorial state" to a "population state"?' (Foucault 2004a: 373; cited in Elden 2007: 563). Or, as Mitchell Dean puts it:

> modern governmentality can be identified by a particular regime of government that takes as its object 'the population' and is coincident with the emergence of political economy ... Thus government involves the health, welfare, prosperity and happiness of the population. (1999: 19)

In light of this, while the previous chapter looked at two important 'peoplings' of this rehearsal state (the Dalai Lama and TGiE bureaucrats) this chapter turns to a third and vital peopling: the Tibetan population and citizenry in exile.

What follows focuses on three takes at the making of exile Tibetan subjects: the construction of the Tibetan population as an entity to be managed; the development of state-like rights and responsibilities through structured welfare provision; and the discursive and material construction of Tibetan citizenship in exile. Running through each of these sections is a focus on both top-down and bottom-up strategies of governance that follows Foucault's twin concerns with the "'micro-physical"... techniques of individualisation that aim to rule [the body and soul] in a continuous and permanent way' and "'macro-physical" strategies of state knowledge-building and regulation... that deal with people as legal subjects and are concerned with the health and productivity of populations and territories' (Clayton 2000: 318). This idea of simultaneous individualising and totalising forms of power evokes the notion of scripting, which I explore in this chapter in terms of both written copy (the plans and documents through which TGiE envisions this population) and the broader act of scripting: of discursively constructing an ideal population and seeking to regulate individual behaviour to achieve this. Again the concept of legitimacy is key here, as I will argue that when scripts are believed – when individuals buy into this rehearsal of stateness – then the authority that the TGiE claims is granted legitimacy.

Scripting and Managing a Population-in-Waiting

> We count 122,078 in exile; females make 44.2% of our population; 74.4% of our people can read and write; 26% of our people work; only 25% of our people are aged below 15; 44.05% of women aged 25–29 are unmarried; infant mortality rate stands as high as 38.9/1000. (Gyalpo 2004b)

With the shift in the eighteenth century to what Foucault has called governmentality, institutions such as the state have taken a particular interest in the people over whom they govern. Central to this has been 'the identification of the people of the state as a population which was understood as the proper focus of the art of government' (Painter & Jeffrey 2009: 29). The population has thus emerged as an increasingly important 'datum, as a field of intervention and as an objective of governmental techniques' (Foucault 1991: 102) to be regulated and optimised. Given both the TGiE's lack of territorial jurisdiction and the centrality of the diaspora to its raison d'être, the 'creation' and the scripting of a population is of heightened importance in this case. This section explores how the Tibetan population in exile comes to be an object of government and how the relationship between governmentality and this population works in this case of a governing agency lacking jurisdiction over territory.

The Tibetan government prior to 1959 knew relatively little about its population. As noted in Chapter 3, the boundaries of Tibetan territory were never clearly defined, regional identities subsumed a broader Tibetan national identity, and no census had been conducted. The process of taking refuge in exile, however, rendered this part of the Tibetan population highly visible as destitute refugees and as ethnically, culturally and religiously distinct from their hosts. For the first few years in exile it was the Indian government that recorded the number of Tibetans entering India and where they (were) moved to within the state.[1] However, as the community became increasingly established, the exiled administration gradually developed mechanisms to know more about its population, to the extent that it has arguably collected more data about the diaspora than the Tibetan government knew about its population in pre-1959 Tibet. Such knowledge was acquired through a range of ad hoc surveys and registration systems including Settlement Office records, the documentation of 'newcomer' refugees by reception centres in Kathmandu, Delhi and Dharamsala, and the administration of TGiE-issued identity documents (Chief Planning Officer, March 2006). Whilst significant in developing administrative and statistical skills within the community, these registration systems were aimed either at a specific cohort within the diaspora or administered within a particular sector (health care, education, etc.). A milestone in how TGiE has come to 'know' its population came with the first census of the exile Tibetan community conducted in 1998.

As a key instrument of statecraft the census is 'an unusually revealing institution through which to trace the mutual influence of the logic of governmentality and the larger cultural context in which it ... [is] embedded' (Hannah 2000: 222). In this case the Tibetan Demographic Survey (TDS) was conducted under the auspices of TGiE's Planning Commission and was 'carried out systematically and scientifically with due cooperation from the Census Commission of India' (Planning Commission 2004: 7). For the second TDS conducted in 2009 almost 2800 Tibetan enumerators were recruited to undertake the census in 220 locations in India and Nepal and 40 enumeration points outside Asia (Planning Commission 2010).[2] With its standardised census format – conducted every ten years with household and individual data stored in aggregate form – the TDS appears, at face value at least, to be more in line with state-organised population censuses than most refugee surveys. The primary objective of these surveys has not been to assist refugee programmes such as voluntary repatriation or resettlement abroad but, in line with state rationales, to facilitate planning in and for the community: identifying needs on a settlement-by-settlement basis, charting population trends, allocating resources and deriving a baseline from which to measure the effectiveness of development programmes.

However, differences between the TDS and conventional national censuses are also revealing. Firstly, with TGiE's limited judicial powers there is no element of coercion in the Tibetan census: not cooperating in the enumeration is not illegal, and there are no penalties. Secondly, an important additional objective of the TDS is as a training exercise in demographic methods, providing the exiled community with experience needed to conduct such a project in 'Future Tibet', which, in turn, would be 'valuable for defining administrative areas and conducting elections [and] to enable equitable welfare delivery' (Planning Council 1994:

Section 2.6.4.1). Third is the issue of territory. The census is conventionally embedded in notions of bounded territory and entails territorialising strategies. In contrast, the TDS is not the counting of individuals within a contiguous space, but rather is a transnational census in its attempts to enumerate the exile community resident in a number of different states. However, given the challenges resulting from the geographical dispersion and high levels of mobility of the exile community, there are significant blind spots in TGiE's 'vision' of the social body (cf. Starkweather 2009). Participation rates have been significantly lower in communities beyond South Asia, and difficulties have been encountered enumerating 'scattered' Tibetans who live outside the TGiE-run settlements (Secretary, CRO Bangalore, November 2007). Yet despite these limitations, the TDS has, through the enumeration process of entering Tibetan homes and its prerequisite land surveys, rendered the spaces over which the TGiE does have autonomy – the settlements in South Asia – as 'legible' and 'calculable' territories (Hannah 2009; Rose-Redwood 2006). This creation of knowledge about these spaces not only facilitates TGiE's governance of the populations within these exclusively Tibetan spaces, but also, in its simplification and rationalisation of the population in these settlements, is an important expression of statecraft (Scott 1998).

As a number of scholars have argued, the census and the statistics generated from it are therefore techniques that do far more than merely provide information about a population (Murdoch & Ward 1997; Scott 1998; Starkweather 2009). Rather, the presentation of demographic data in the form of tables, graphs and diagrams is a key mechanism through which the population is rendered as an entity to be known and thus made 'amenable to intervention and regulation' (Miller & Rose 1990: 5). In the case of the TGiE the administration claims that through the TDS it knows 'the quality and constitution of the Community's human resource base' (Planning Commission 2004: 18) and has developed a range of strategies for examining, problematising and rationalising its internal dynamics. Encapsulated in the 'problem tree' produced for the third Integrated Development Plan (IDP; Figure 5), and echoing Foucault's understanding of problematisation as political governmentality becoming a 'problem that affects those governing and those being governed', a number of 'critical demographic and socio-economic issues' (Planning Commission 2004: 18) are perceived to compromise the ongoing success of this exile population. The laying out – and one might say totalising – of these policy issues has the effect of generating a demand for long-term planning and intervention.

Framed as key to improving the lives of the exiled population, an infrastructure of centralised planning was institutionalised through the establishment of the Planning Council in 1988 (renamed as the Planning Commission in 2003). Modelled on the Indian Planning Commission's Five-Year, this body has, to date, formulated four five-year IDPs that set out the development priorities for the community and outline individual projects for prospective funders.[3] In thus both offering an overview of the exile community's attributes, needs and

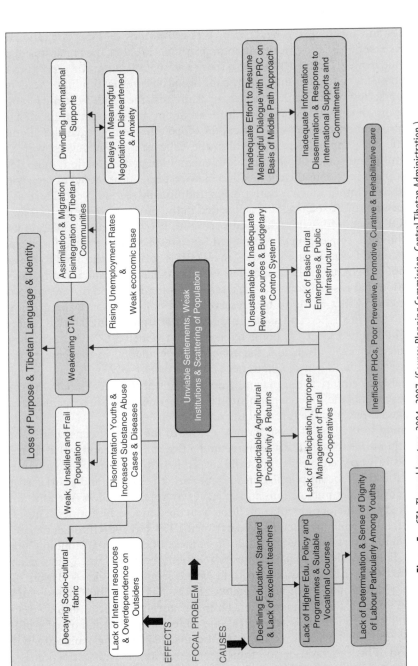

Figure 5 CTA: The problem tree, 2004–2007. (Source: Planning Commission, Central Tibetan Administration.)

priorities and meeting 'the demand of Western friends of Tibet for rationality, objective standards, and prioritization of projects – in short, the apparatus of efficient modernity' (Lafitte 1999: 158), these national plans can be seen as examples of nascent state planning. Albeit contingent on external funding, these 'road maps' or 'scripts' for the exile community distinguish the TGiE from most refugee or diaspora organisations and constitute an important articulation of stateness. They also demonstrate how the situation of exile has encouraged the adoption of a different temporal ontology within the Tibetan bureaucracy as, with their goal-setting and stages of planning, implementation and evaluation, these development plans represent the codification of a Western, linear model of time within a Buddhist community for whom time is traditionally seen in cyclical terms.

Returning to the societal 'ills' delineated in the problem tree produced for the third IDP (Figure 5), whilst the TGiE seeks to 'know' its exile population using the techniques conventionally employed by states, how this population is imagined, planned for and understood by the TGiE is specifically attuned to the exile situation. In light of the threatened loss of Tibetan language and identity, this is a population that is scripted as having a series of interlinked and distinct purposes: as a 'resource' that needs to be preserved; as a population-in-waiting ready to return to a future Tibet; and as a cultural repository, preserving a unified Tibetan national identity outside the home territory (McConnell 2012). As such, the circumstances of exile mean that the concept of population in this case carries a 'normative burden': contra Chatterjee (2004) it is not *solely* a domain of policy, identifiable by empirical criteria and amenable to statistical techniques. Through specific scripting the Tibetan population is also, at one level, a homogenising exercise linked directly to the future of the nation.

Turning to the first of these scripts, central to the TGiE's construction of its population is the idea of it as a resource under threat in Tibet and which therefore needs to be preserved in exile. The exile government's concerns with demographic changes in the Tibetan population in Tibet are a recurrent and politicised feature of government publications and speeches by political leaders (see Shiromany 1998) and are a major point of contention with the Chinese authorities. The TGiE alleges that 1.2 million Tibetans died between 1959 and 1986 through detention, famine and poverty as a direct result of Chinese occupation (Planning Council 2000). This is a figure that, despite its reliability being disputed, 'has become enshrined as an incontrovertible truth in exile discourse' (Childs & Barkin 2006: 40). In addition, the TGiE and Tibetan NGOs claim that the Chinese authorities are undertaking 'demographic aggression' through policies of forced abortions and sterilisations in Tibetan areas and deliberate population transfer of Han Chinese into Tibet, with the result that Tibetans are becoming a minority in their own land. This portrayal of an ethnic group threatened with extinction is discursively linked by the TGiE to the vitality of the exiled population, with the construction of the latter as a population whose numbers need to be preserved. In light of such anxieties, and following the publication of the 2009 TDS results, a key concern is the low and declining fertility rate – from 4.9 births per woman (of child-bearing age) in 1987–89 to 1.18 in 2009 – and the subsequent 'threat to the sustenance of the Tibetan community in India ... and erosion of [our] very purpose' (Planning Commission 2004: 30; see also Childs 2008).

This discourse of the Tibetan population as a resource has important resonances with state-like strategies of biopower. Not only has the TGiE established as one of its central concerns the care, health and reproduction of the exile population, but it is also seeking to optimise the productivity of this population in order to fulfil its broader political project. As one would expect, the 'cast' in this rehearsal of stateness is seen to play a fundamental role. Central to this is the discursive construction of this population as 'in waiting', ready to return to the homeland in the future. As explored in the previous chapter, a key element of this script is the training of exile Tibetans in a range of skills of governance, from development planning and demography to formulating legislation and auditing accounts in order to implement these practices in a future Tibet.

However, though important, discourses around biopolitics tell only half the story. Alongside the vitality of the Tibetan population being under threat from Chinese population policies, Tibetan identity and culture are also perceived as endangered. As a result, and articulated as one of the main reasons for coming into and remaining in exile, the diaspora is perceived as a cultural and national repository. This is clearly set out in an article in the TGiE's *Tibetan Bulletin*:

> The purpose of the Tibetans in exile is two-fold, viz., to seek justice for our homeland and to preserve our identity and language by practicing our culture and traditions. The first purpose is dependent on many factors including international situation, political changes within China etc. that are beyond our control ... However, the second purpose is not dependent on external factors and can be fulfilled by every Tibetan in exile. (May-June 2003: 16)

Underpinning this project of preserving identity and culture in exile is the perceived need to foster a very particular kind of population in exile: a cohesive, united and homogeneous community that shares a single national identity.

With the script having been written, the task has been to persuade the cast to believe in it. To this end the TGiE has devised a range of strategies to seek to regulate individual conduct in order to achieve this 'ideal' of a healthy, united and nationalised exile population. Attending first to discourses regarding the Tibetan population as a scarce resource, the TGiE employs regulatory mechanisms of biopolitics through seeking to manage the health and reproductive behaviour of its exile population. The TGiE's Department of Health, established in 1981 to 'plan a comprehensive health care system for the Tibetan Refugee Community' (Planning Council 1994: Section 6), currently administers an extensive network of hospitals, primary health centres and clinics in the exile settlements. Alongside these medical institutions are a range of public health initiatives, with settlement-based programmes for disease control, health education and improving sanitation, and access to drinking water.

In terms of promoting the reproduction of the population, mother and child health and have been initiated by the TGiE and rolled out across the settlements, and stable family units – or, as they are termed in the TDS, 'normal household situations' (Planning Council 2000: 12) – are promoted through the allocation of settlement houses. More broadly, and echoing a well-rehearsed relationship between reproductive practices and national progress (Hodges 2004), pronatal-ism and ethnic endogamy are promoted by the TGiE as essential to stemming the perceived threat to Tibetans as a distinct ethnic group. As such, members of the exile community are 'encouraged to further a nationalistic agenda, promulgated and disseminated by their leaders, by reproducing exclusively with Tibetans at a sufficient rate to ensure population growth' (Childs & Barkin 2006: 49). How-ever, the translation of such policies into practice has, predictably, been some-what limited. Whilst intermarriage between exile Tibetans and members of their South Asian host communities is relatively rare, marriages with 'Westerners' do not receive the same levels of social reprobation. In terms of pronatalism, though advocated through TGiE campaigns and publications, low and declining fertility levels indicate that factors such as high female non-marriage rates, increasing lev-els of education among Tibetan women, and the costs of raising children often outweigh any pressures from nationalist ideologies (Tibetan Women's Association 2005).

Although in the case of endogamy and pronatalism the TGiE's script appears to be falling short in both matching the realities in the community and convincing individuals to adhere to it, the administration's regulatory strategies to preserve and promote Tibetan cultural and national identity have seen considerably more success. As a young teacher in Sonamling settlement, Ladakh, explained:

> The power and authority of our government, it comes from sentiments ... patriotic and heartfelt sentiments of the people – what is in their hearts not practicalities of our government because its finances aren't big at all ... it's because of people's nationalism that it exists and is so important (May 2007).

It is precisely through this construction of nationalism that the TGiE is most effective in asserting its authority: an authority of 'affective ties, moral claims and authentic arrangements' (Feldman 2008: 15) that lie at the heart of the mod-ern nation-state. At first glance the TGiE's strategy of subsuming older regional and sectarian identities under a homogeneous Tibetan nationalism that had pre-viously not existed beyond the Lhasa elite has had mixed results. In many ways regional and sectarian divides have not disappeared, and many of those I spoke to cited the parliamentary electoral system as having exacerbated factionalism in exile, as well as the dominance of mutual aid groups known as *kiduks*, which are conventionally organised by an individual's native area (*pha yul*).[4] Yet, at the same time, most interviewees – especially those from the younger generation – also spoke of 'Tibetanness' being their most important identity marker, reflecting Emily Yeh's observation that regional identities have been 'largely papered over in the transnational nation-building' (2007: 650).

The reinterpretation and in many cases invention of a range of 'national' traditions has been key to constructing a powerful imagined sense of solidarity and belonging. These patriotic yet banal acts (Billig 1995) include flying the national flag, staging debates on the legitimacy of Tibet's claim to independence, and participation in national holidays such as the anniversary of the national uprising on 10 March 1959. Such performances of Tibetan nationhood therefore perpetuate the ideal of national unity and instil a collective sense of Tibetan identity. Tallying with scholarship in anthropology on the essentialisation of national identities by state elites (Gupta & Ferguson 1992; Hobsbawm & Ranger 1983), it is the Dalai Lama and the TGiE leadership who are the 'primary authors of these reworkings' (Houston & Wright 2003: 222). This is a role that, in turn, reinforces the administration's legitimacy:

> We are struggling to create Tibet outside Tibet... and for this our exile government provides guidance – it plays key roles in making Tibetan identity strong with promotion of culture, language and Tibetan way of life... The government and its schools and institutions keep the Tibetans as Tibetans. (Former Education *Kalon*, December 2006)

Another key arena in which Tibetan identity is standardised and regulated is that of education. It is widely accepted that school curricula construct a citizen's world-view and sense of national identity (e.g. Regalsky & Laurie 2007). In the Tibetan case education has been accorded the highest priority since the early years of exile, and the Department of Education currently oversees over 70 schools in India and Nepal, serving around 24,000 Tibetan children. Alongside adhering to the Indian government-approved syllabus, a 'curriculum of Tibetanness' (Kolås 1996: 57) is imparted through classes on Tibetan language, culture and religion, taught by Tibetan teachers using textbooks published by the TGiE's Department of Education. In documenting the role of exile education in promoting an essentialised notion of Tibetan nationalism, Margaret Nowak noted that this was a 'systematic attempt to shape a more cohesive group identity in exile' (1978: 71) especially amongst second-generation exiles. Such is the pivotal role of education in the eyes of the TGiE that failings, particularly around issues of language (see Figure 5), are perceived to jeopardise the broader project of preserving Tibetan culture and identity. In light of such concerns, the TGiE launched a 'Basic Education Policy' in 2005, which (re)emphasises traditional Tibetan values taught in Tibetan language (Department of Education 2005). As such, this policy aims to produce a very specific and idealised 'type' of

Tibetan – traditional and modern, non-violent and truthful – through what are in effect a series of disciplinary institutions (schools) and technologies (curricula). The policy has, however, proved controversial, with concerns that Tibetan-medium education disadvantages Tibetan students in the Indian job market or looking for careers in the West.

While the Tibetanisation of the school curriculum is a (contested) strategy aimed at preserving national culture and identity, perceived threats to the cohesiveness of the community are addressed by identifying a series of distinct cohorts within the exile population who are failing or refusing to remain within the established Tibetan spaces and communities in India. Once identified these cohorts are, reflecting the techniques of governmentality, then subject to distinct policies (Chatterjee 2004).The first cohort is the thousands of refugees who, as a result of resource constraints, remain 'unsettled' in India and Nepal and thus 'do not belong to a viable Tibetan community in which they can preserve their language and culture and give their children a Tibetan education' (Planning Council 1994: Section 3B.1.1). Prompting similar concerns are those individuals who leave the settlements in the winter months to sell sweaters in cities across India. Although this is one of the earliest trades that Tibetans engaged with in India, and is still a mainstay of many communities, the social effects of this annual migration, especially on the institution of the family, are viewed as highly problematic. As the Additional Secretary at the Department of Home explained:

> this seasonal work, it affects the sustainability of settlement. Able-bodied adults are away and ... how to say ... family atmosphere is not there. This doesn't serve our main aim of settlements, of keeping the community together (November 2007).

Such concerns are also articulated with regard to a third cohort: youth. Frequently mentioned in TGiE policy documents and a recurrent topic of conversation in my interviews, young Tibetans educated in exile are more exposed to Indian and Western society than are the older generations, prompting concerns regarding loss of culture, antisocial behaviour and increasing numbers emigrating to the West.

Having identified these 'wayward' cohorts and constructed them as problematic, the TGiE has sought to regulate their behaviour with the goal of integrating them into the 'mainstream' exile community. Central to this have been attempts to control their residency and mobility, from assigning vacant houses within settlements to 'unsettled' Tibetans to encouraging sweater sellers and young Tibetans residing in Indian cities to 'resettle' (Secretary, Department of Home, November 2007). In replicating the quintessential state project of seeking to settle mobile peoples these policies carry with them connotations of social regulation (Scott 1998). Exile officials frequently spoke of establishing a desired order and stability and of persuading 'troublesome' populations away from the perceived social dangers they are exposed to outside the community. Echoing the fact that state policies of sedentarisation so seldom succeed (Scott 1998), these regulatory policies have faced significant challenges and limitations, both from the constraints of operating within a host state and resistance from the cohorts at which they are aimed. Although its significant moral authority goes some way to explaining the compliance of Tibetans with TGiE settlement policies in the past, the TGiE ultimately lacks the legal authority and coercive powers to determine the movement of its exiled citizens. Moreover, in a number of cases the TGiE's policies of settling scattered Tibetans have not been received in the way that they were intended. For example, several interviewees in Dekyiling spoke of families who were granted a house in the settlement continuing to trade in the hill stations in Uttarakhand, renting out their settlement house or using it for storage. Such (ab)use of the system therefore indicates some of the limitations of TGiE's governance, and perhaps an increasingly cynical attitude to state-citizen relations (cf. Navaro-Yashin 2006).

However, such limitations of the TGiE's regulatory mechanisms appear almost insignificant when compared to perhaps the most problematic demographic issue that the community faces: the division between Tibetans in Tibet and those in exile. It is imperative to remember that this construction of a nationalised population has generated a 'new' national body formed primarily in exile. As such, whilst there is often regular contact between family members in Tibet and those in exile, and between monasteries in Tibet and their re-established institutions in India, formal communication between Tibetans in Tibet and the exile government is both logistically difficult and politically dangerous for the former. As noted in Chapter 4, the exile government has sought to symbolically bridge this divide through the quota system for parliamentary elections but this symbolism appears tokenistic when faced with the task of uniting populations that have been divided by the Himalayas for over 50 years. The decades-long project of cultural preservation in exile has arguably produced both a static and a conservative version of Tibetan culture, and one increasingly influenced by Indian and Western cultural traditions (Yeh & Lama 2006). As Emily Yeh has noted, cultural differences between newcomer refugees and those brought up in exile have led to tensions over issues of identity and authenticity:

> the fact that Tibetan identity in exile has been constituted in opposition to China contributes to both the scorn and suspicion of the 'Chinese' appearance and behavior of new arrivals who, because they are different, are seen as less than authentically Tibetan … as a result, many new arrivals report that they feel like outsiders among Tibetan exiles in India. (2007: 653–654)

While cultural divides between Tibet and exile are starting to be slowly bridged, in part due to the rise of the 'Lhakar' cultural movement (McConnell 2015) and the cross-border popularity of pop music from Tibet, nevertheless social marginalisation of refugees newly arrived from Tibet continues. As noted in the previous chapter, the relative dearth of Tibetans from Tibet working in the TGiE and elected as *Chitues* perpetuates this marginalisation and has led to criticisms from some newcomers that 'our exile government has become a government for Tibetans born in exile, not for those of us from Tibet' (monk recently from Tibet, March 2007).

Focusing on the TGiE's construction of and relationship with 'its population' is therefore an important reminder that this project of rehearsal is an *exile* project: one that is time and space specific, and where threats to the Tibetan population and culture in the homeland have shaped the discourses through which this population is imagined and idealised. Yet, within these constraints, the TGiE has been broadly successful in constructing the Tibetan population in exile as an entity over which it has responsibility and thus as an object of government. In partitioning out and attempting to manage this domain of governance, these are strategies that go some way to legitimising TGiE as a governing authority. For, without territory to legally call its own, it is the population that has followed the Dalai Lama into exile that has validated the exile administration's state-like existence and continued functioning. As outlined here, the TGiE has strategically employed a range of state-like techniques of governmentality – from carrying out a national census to implementing school curricula, seeking to settle 'wayward' cohorts and promoting particular reproductive practices – in order to act like a state for domestic and external audiences. The scripts have been successfully drafted, and the project of cajoling the cast into engaging with the performance is a work in progress.

Welfare Provision and Delineating a 'Civil Society' and 'Economy' in Exile

Though revealing in terms of the state-like scripts through which the TGiE observes and regulates those that it seeks to govern, population management is but one aspect of the construction of Tibetan subjects. Following Colin Flint's assertion that the state must be defined 'in relation to two other spheres of modern life, the market... and civil society' (2009: 723), I want to turn here to the ideologies, institutions and practices of welfare, and the construction of civil society and a nascent economy in exile. In terms of providing for the needs of a population, welfare is a key technique of governmentality through which life is managed and governed. However, in acknowledging that 'regulation' is perhaps not the best word for actions undertaken to enhance social welfare, Matthew Hannah (2000) points to the fact that welfare also encompasses important political ideas and ideologies: of rights and responsibilities and a political contract between the citizen and the state. As anthropologists of the state have asserted, it is through everyday exchanges regarding health care, entitlements, education provision and sanitation facilities that individuals – and in particular poorer individuals – experience the state most directly (Das & Poole 2004; Hansen & Stepputat 2001). Keeping this in mind, it is the relationship between welfare and the state that is of primary concern here, a relationship of which there are conflicting accounts. On the one hand, welfare is seen as a basic state function, with the state having a moral obligation to look after its people (Taylor 1994). In contrast to this 'welfare state' model, and in light of increasing devolvement of welfare provision to non-state actors under neoliberal policies, is the development of the idea of a 'welfare society' (Gould 1993). No longer an essential component of statehood, welfare in this reading is provided by a combination of voluntary organisations, private corporations and quasi-state agencies. As I explore in this section, it is through the TGiE's own provision of welfare and through its negotiating relationships with non-state welfare providers that we can catch glimpses of the TGiE's rehearsal of statehood.

Attending to the welfare needs of Tibetan refugees in South Asia has been a principal aim of the TGiE since its establishment in 1960. Over the decades a series of welfare programmes have been institutionalised, to the extent that Tibetans in India today benefit 'from an almost cradle to grave welfare system' (Klieger 1992: 102). At the core of this welfare provision is the Department of Home, which has the responsibility to 'look after the socio-economic welfare of the Tibetan refugees in exile ... so as to achieve the long term goal of self-sufficiency' (www.tibet.net). As noted in Chapter 4, the expansion and standardisation of welfare facilities across the settlements can be regarded not only as a strategy for integrating exile Tibetan society but also as an attempt by the TGiE to extend its responsibility for its citizens as comprehensively as possible. This idea of welfare provision as a duty and responsibility that the exile government has to 'their people' is articulated by Tibetan officials at a range of levels. For example, according to the Secretary of the Department of Home, he is 'mandated to look after the socio-economic welfare and development of the Tibetan community in exile' (November 2007), while at a local level the Settlement Officer at Dekyiling explained that 'our settlement people must be settled down ... all facilities they must be provided and our people must face no problems living here. This is our important duty' (April 2007). Such rhetoric of responsibility, duty and an ethic of care, along with the state-like connotations of the repeated use of 'welfare' and 'entitlements' rather than terms such as 'charity' or 'aid' indicates that, from the TGiE perspective at least, this is a state-citizen rather than NGO-recipient relationship. Looking at the story 'from below', many interviewees in the settlements, whilst not expressing it in the language of 'rights' per se, did generally *expect* their government to provide basic welfare services. Respondents were, in the main, impressed by what the exile government had achieved, appreciating that their access to and quality of education and health care was, in many cases, superior to their Indian counterparts and to co-patriots in Tibet.

In addition to attempting to provide standardised welfare provision, the TGiE also engages in a process of individualisation through identifying, problematising and then (differentially) regulating cohorts within the exile community. Perhaps the most important of these cohorts in welfare terms is 'the poor'. In recent years the exile government has invested considerable effort in the 'production of persons who can be labelled as poor' (Corbridge et al. 2005: 47). At the 'national' level the TGiE established a 'poverty line' in 2002 of 30 Indian rupees per person per day to assess poverty levels within the community (since updated to 1 US dollar a day; Tenzin Sherab 2012: 22), and formed an inter-departmental 'Central Poverty Alleviation Committee' to 'identify poor persons, so as to formulate suitable strategies for intervention and so achieve upliftment of these families from the poverty line' (Secretary, Department of Home, November 2007). At a local level Settlement Officers instruct the *Gyabons* to conduct regular door-to-door surveys in their camps to 'see the local conditions and report problem cases which can then investigate and get timely welfare support for that person' (Settlement Officer, Sonamling, May 2007). Once thus identified, TGiE departments, CROs and Settlement/Welfare Offices implement a series of welfare assistance programmes including monthly stipends for the elderly, school scholarships, grants to cover medical expenses, business grants and vocational training. This is therefore a process of identification, problematisation and regulation through which poverty is constructed as a concern that is both concentrated upon certain social groups, and that affects the wider body politic.

However, whilst the TGiE aspires to a universal, state-like welfare system within exile, this project is hampered by the administration's lack of jurisdiction over bounded territory. The result is that welfare provision is restricted to Tibetans residing in India (and to a lesser extent Nepal) and within this section of the population, access to welfare services is highly uneven, being disproportionately concentrated in exile institutions such as monasteries, schools and refugee reception centres. Not only do around a quarter of the exiled Tibetan population live in residential institutions (Planning Council 2000) but, rather than 'demarcated spaces to which ... socially dependent populations have been more or less forcefully "exiled"' (Philo & Parr 2000: 513) these institutions are a central and vibrant part of the community. As well as being a key element in the project of cultural preservation, this archipelago of Tibetan institutions across India is framed by the TGiE as encouraging Tibetans to lead lives of 'benefit to society' (General Manager, Rajpur handicraft settlement, April 2007) and constitute the spatial concentration of welfare and governmental dependency.

Illustrating this has been the example of new arrivals from Tibet who use a well-established route into exile via the TGiE's refugee reception centres in Kathmandu, Delhi and Dharamsala. After medical checks, testimonial interviews and an audience with the Dalai Lama, the Dharamsala centre seeks admission for these individuals into a range of exile institutions 'to prepare for their long-term rehabilitation' (Director, Dharamsala Refugee Reception Centre, April 2007). Children up to the age of 18 are sent to Tibetan residential schools, former political prisoners to the Department of Health's 'Torture Survivors Programme', monks and nuns to monasteries and nunneries across India, the elderly to old people's homes, adults aged 18–30 to the 'Tibetan Transit School', and older lay adults to settlement-based handicraft centres. This series of institutions is therefore an important mechanism through which TGiE attempts to incorporate newcomers into the community and provide vital welfare support. As such, these refugees are often appreciative of and dependent on the exile government. However, after individuals graduate from or leave these institutions they often face significant problems. With little or no savings, often no family in exile and a lack of Indian-recognised qualifications, newcomer refugees frequently spoke of their insecurity and vulnerability within India and difficulties integrating into the exile community.

In presenting two contrasting pictures of the Tibetan 'welfare state' in exile – one of a state-like welfare ethos seeking to roll out standardised support; the other of uneven access to services with provision existing primarily in Tibetan institutions and settlements – I want to consider what this means for the TGiE's attempts to construct itself as a government. For many Tibetans resident in settlements or institutions the existence of the exile government was a source of comfort and even security in the otherwise vulnerable situation of exile:

> As we are a homeless nation our government is the only real thing we have … that we can call our own. So even if our government is not recognised by any other government, still we are so lucky to have it. Because of [our] government we don't feel like refugees – everything is provided for us … all is organised by Tibetans. (Monk, Lugsum-Samdupling, November 2007)

However, a number of government interviewees expressed frustration that they could not do more for 'their people'. As one *Chitue* explained when describing his visit to Tibetan settlements in Arunachal Pradesh:

> I was amazing [sic] that these remote places have so many Tibetan structures … they were so far away but so familiar and all established by our government. But then also I feel frustrated that we cannot do more. I feel we should be doing more for our people, providing more, helping them more, but always we come up against problems – against barriers because we are in India. We can do little since we do not have the stamp of being a state … we cannot be a complete government for our people here (September 2011).

This perception that, in its failing to provide for the exile population the TGiE is falling short of its 'government' title, was echoed in a discussion between an informal social worker (Thupten) and older shopkeeper (Jampa) in Majnuka Tilla (in June 2007):

> *Thupten:* For me I think our government could do more for our people... in some remote settlements – not Bylakuppe but in North-East – there people are so poor and have so much hardship. Our government it should be working to help them for day-to-day things, but instead all we have is meetings and discussion.

> *Jampa:* ... but I think our government, it does OK, and we should not criticise. We are refugees, we have no country and here we are on borrowed land, so how can our government do all these things?

> *Thupten:* [interrupts] but living on borrowed land does not mean the responsibilities they are not there. We have no country but still our government it has to work. Why do we have our government here in first place? If our government does not help its people then why is it a government?

> *Jampa:* But for our government its hands are tied because we are in India. We have no land, we have not so many rights – the Indians they can tell our government you can do this, or you can't do this. But at the same time our youths they expect the government to help them with everything – with education, with work... the youths they are too dependent on our government, and on help from foreigners. For our people we need more initiative.[5]

Such snapshots of opinion portray complex and conflicting accounts of the TGiE. On the one hand, the TGiE cannot meet both its own and its citizens' expectations of a state-like provision of welfare. But, alongside this assertion that there is 'not enough' government, Jampa's comments also suggest a criticism of there being 'too much' government. Indeed, this last exchange in many ways rehearses political debates familiar in established democracies. However, Jampa's argument also points to an important additional viewpoint, one that is specific to the situation in exile and questions the very purpose of the TGiE. This is the sense of frustration from some within the community that the TGiE has over-prioritised the welfare needs of the exile population at the expense of the freedom struggle and the ultimate aim of returning to Tibet. In this case the rehearsal of stateness is perceived to divert attention from more pressing political imperatives. The fact that the TGiE's 'welfare state' aspirations are themselves controversial means that belief in the script of a welfare state in exile, with the attendant rights, duties and obligations, cannot be taken for granted. Given the precariousness of exile, the TGiE must work hard to position itself as the core and 'natural' provider for its population as it does not have a monopoly on welfare provision. An array of different actors – from international donors upon which the TGiE's welfare programmes are heavily dependent, to the Indian government and Tibetan NGOs – vie for responsibility for the exile population, and encroach on the space carved out by the TGiE's 'welfare state'.

The Indian authorities were instrumental not only in establishing the Tibetan settlements but also in providing basic housing and rations in the early years (TPPRC 2006: 116). Whilst there has been a gradual transition of administrative authority and responsibility from the Indian government to the TGiE, there continue to be fuzzy boundaries in terms of welfare service provision, with Tibetan health and education facilities open to local Indians, and Indian schools and clinics available to Tibetans. Meanwhile within the exile community there are a plethora of welfare-focused Tibetan NGOs. These include: Tibetan Children's Villages, which run schools and higher education programmes for almost 17,000 children in India; Kunphen drugs rehabilitation centre in Dharamsala; GuChuSum, which provides welfare for former political prisoners; and a series of *kiduks* which provide financial and social support for their members. Whilst this provision of welfare support by Tibetan NGOs is accepted without question by the TGiE, the relationship between the exile government and these organisations came under scrutiny during Samdhong Rinpoche's administration in ways that are revealing of the TGiE's construction of itself as a government. For many years Tibetan NGOs have had to formally register with the TGiE,[6] but a series of policy directives issued by the *Kashag* in July 2007 set out the relationship in stricter terms:

> NGOs should be non-governmental ... the NGOs should not ... rely on or look up at the government. Neither should the government ... interfere in, exercise influences over or otherwise make use of NGOs. Thus, it is important for both sides to maintain a proper standard of relationship and distance on a stable basis. (*Tibetan Review* 2008: 26)[7]

Such a distinction between the TGiE and Tibetan NGOs is based on the rationale that, with the TGiE's pursuit of a policy of autonomy in Tibet, it was necessary to distance itself from NGOs that have 'independence agendas'. The establishment of boundaries for NGOs and the regulation of government interaction with them thus enables the TGiE to distinguish itself from the very type of institution that it has sometimes been labelled as, and instead position itself within discourses of stateness. This delineation can be read as part of a wider TGiE project of creating a sphere of Tibetan 'civil society' in exile distinct from itself as a government. By this I am not invoking a notion of civil society either in an expanded sense of all social institutions that lie beyond the domain of the state, nor a narrow sense in terms of 'the closed association of modern elite groups' (Chatterjee 2004: 2). Rather I am interested in the TGiE's demarcation of specific institutions and activities as lying outside of its defined remit. Thus, read alongside Timothy Mitchell's (1991) assertion that, rather than the separation of state and civil society underpinning the organisation of power, the state is a set of effects that *make* structures such as the state and society appear to exist, the TGiE's attempted construction of these separate spheres can be seen as an important legitimising strategy.

However, it is not just exile civil society that the TGiE is seeking to distinguish itself from. There are also moves to separate this polity from a nascent exile economy. At first glance the limitations facing TGiE's control over the economic practices of Tibetans in South Asia appear considerable, and the idea of an exile Tibetan 'economy' somewhat fantastical. For a start, its lack of bounded territory means that its ability to define, foster and regulate a 'national' economy is severely restricted and its job creation capabilities very limited. Moreover, given its existence within India, the TGiE has no currency, no national bank and therefore no monetary policies. Tibetan economic activities are both highly dependent on and restricted by the Indian state, with individual Tibetans taking loans from Indian banks, Tibetan businesses having to register with the Indian authorities, and wealthier Tibetans paying income tax to the Indian government. Economic dependency is also evident in the TGiE's ongoing relationship with international donors. The administration's welfare programmes are heavily reliant on external funding from international NGOs and charitable trusts, and foreign patrons are also sought for various exile Tibetan institutions.[8] At the individual level, a widespread system of sponsorship of Tibetans by Western benefactors (*rogs ram*) continues to play a significant role in Tibetans' livelihoods (Prost 2006).

Yet despite these limitations to economic autonomy, a recurrent narrative in interviews with TGiE officials was a series of developmental transitions that the government had overseen during its decades in exile, from meeting basic welfare needs, through the rehabilitation of Tibetan society within the exile settlements, to contemporary aspirations of economic sustainability. In the early days of exile the upheaval of displacement and the re-establishing of a society and government in a foreign land inevitably brought with it rapid transformations in economic practices and relations. Dispossessed of their land and the taxes it accrued, monasteries and aristocratic landlords were no longer the economic elite. The reliance instead on settled agriculture, petty business and tourism thus 'generated new relations of production and exchange, both communal and individual, and new class, gender and ethnic structures' (McGukin 1997: 237). Such fundamental shifts forced the Tibetan leadership to reconsider the economic relations it had with its now exile population, but it is only in recent years that economic issues have come to be seen as 'a catalogue of problems *for* government' (Murdoch & Ward 1997: 310, emphasis in original).

Problems of youth employment, sustainable livelihoods, financial security and the continuing dependence on foreign aid are seen by the TGiE to threaten the idealised vision of a cohesive and united exile population. Consequently, and echoing Mitchell Dean's observation that '[t]o govern properly, to ensure the happiness and prosperity of the population, it is necessary to govern through a particular register, that of the *economy*' (1999: 19, emphasis in original), the realisation of a financially self-reliant administration and economically self-sufficient settlements has become a policy priority of the administration. Documented in the recent IDPs, statements by the *Kashag* and *Sikyong*, and a USAID-funded survey of economic development programmes (TechnoServe 2010), these goals are being pursued through the diversification of agriculture and the establishment of small-scale industries within the settlements, and the promotion of private enterprise through micro-credit provision and training for young entrepreneurs.

The TGiE's relationship to business more generally is revealing of how it seeks to express its stateness. In recent years the exile government has both distanced itself from and formalised its relationship with private enterprise in exile. At a 'national' level, the TGiE for many years ran a number of businesses in Nepal and India, including hotels, a travel agency, gas distribution company and agricultural feed business, the profits from which contributed directly to the TGiE budget. However, declining profits in the 1990s and the *satyagraha* political philosophy of Samdhong Rinpoche's administration (based on truth, non-violence and democracy) saw the 12th *Kashag* privatise these businesses and return the capital deposits totalling 211.7 million rupees to the Tibetan public (Paljor Bulletin 2005: 22). Following this privatisation the TGiE prioritised the promotion of trade and commerce *within* the community, framing this as key to providing employment opportunities, fostering links across the diaspora and diversifying the exile economy (Business Officer, Department of Finance, April 2006). Yet, whilst the TGiE actively promoted private enterprise in the community it simultaneously distanced itself from this newly supported sphere of Tibetan business. As Samdhong Rinpoche outlined in a statement to the Tibetan Chamber of Commerce:

> Generally, a government is best who governs the least. That is why even TGiE is trying to engage itself the least in the businesses of Tibetan people. Therefore ... from our side, it is important for TCC to stay as a NGO in principle without having to depend on TGiE and should formulate its own plans and policies. (TCC 2006: 9)

Such 'disengaged engagement' with the exile business community is, I want to suggest, the TGiE negotiating between intervening in the lives of its citizens, and retracting from such intervention: a delicate balance between state regulation and pulling back from governing in certain areas to allow the 'free enterprise of individuals' (Rose & Miller 1992). This attitude of managerial liberalism can, therefore, be seen as the TGiE experimenting with a more mature stage of governmentality. Following Cynthia Weber's observation that, in order to retain control, authorities simulate boundaries that mark the range of their 'legitimate powers and competencies' (1995: 129), the exile government is thus seeking to fashion a Tibetan population, civil society and nascent exile economy as quasi-objects separate from the 'political' realm through which they are governed. However, though a focus on the construction of such abstract spheres of rule is revealing in terms of the TGiE's expression of and aspirations towards state-like governance, there is a danger of losing sight of exile Tibetans themselves and of their relationship to the exile administration. In asking how belief in the TGiE's scripts of governance is engendered, how Tibetan subjectivity has been constructed in exile, and what the nature of the social contract is between the TGiE and individual Tibetans, I now turn to the question of citizenship.

Collective Scripts of Citizenship *and* Refugeehood

At the heart of the relationship between political identities and the modern state is the construction of a binary between the citizen resident in a bounded, national community and its archetypal 'other', the refugee. The state conventionally employs this binary to shape political subjectivities and to define both the boundaries of inclusion into and exclusion from the national body politic (Agnew 1999). The fact that exile Tibetans are simultaneously 'Tibetan citizens' in the eyes of the TGiE, 'refugees' in the eyes of many within the international community, and 'foreign guests' in the eyes of the Indian state therefore fundamentally disrupts this conventional mapping of citizen and refugee onto concepts of statehood and statelessness (see McConnell 2013a). More than this, the TGiE itself actively promotes *both* Tibetan citizenship and refugeehood as central to exile political subjectivities, thereby articulating dualistic state/non-state relations. In returning to the TGiE's construction of and claims to legitimacy, it is the scripts and practices of *identification* through which notions of citizenship and refugeehood are constructed that is of particular interest here. A processual term that focuses attention on the agents who do the identifying, 'identification' highlights how the state 'monopolizes, or seeks to monopolize, not only legitimate physical force but also legitimate symbolic force ... [which] includes the power to name, to identify, to categorize, to state what is what and who is who' (Brubaker & Cooper 2000: 15).

Turning first to citizenship, Tibet, as it was governed under the Dalai Lamas prior to 1959, had neither a single category of the citizen nor the conferment of homogeneous rights across the Tibetan population (Frechette 2006). Rather, the relationship between the Tibetan state and its inhabitants was, in very general terms, contingent on an individual's landholdings and position within socioeconomic hierarchies (Goldstein 1989). The TGiE, however, has developed a formalised notion of Tibetan citizenship, the definitions, criteria, rights and duties of which are enshrined in the 1991 Charter. Tibetan citizenship is granted thus: 'All Tibetans born within the territory of Tibet and those born in other countries shall be eligible to be citizens of Tibet. Any person whose biological mother or biological father is of Tibetan descent has the right to become a citizen of Tibet' (Article 8). The rights of such Tibetan citizens include equality before the law

(Article 9), religious freedom (Article 10) and freedom of movement and association (Article 12). Under Article 13 of the Charter:

all Tibetan citizens shall fulfil the following obligations:

(a) bear true allegiance to Tibet;

(b) faithfully comply and observe the Charter and the laws enshrined therein;

(c) endeavor to achieve the common goal of Tibet;

(d) pay taxes imposed in accordance with the laws;

(e) perform such obligations as may be imposed by law in the event of a threat to the interest of Tibet.

These statutes of Tibetan citizenship appear, at first glance, to represent the endpoint in a familiar transition in modern state relations from submissive subjects whose obligations are based on their position within set hierarchies to one grounded in citizenship and the language of rights and social responsibilities (Miller & Rose 1990). As such, this scripting of citizenship forms a key element both in the broader exilic nation-building project and in the TGiE's efforts to emulate liberal democratic governance. As Frank Cassidy and Robert Bish note in their study of first nations governments in North America, the 'authority to define the criteria, benefits and responsibilities of citizenship is one of the most treasured of jurisdictional powers claimed by sovereign governments' (1989: 53) and is therefore an authority that non-state aspirants often seek to claim. However, exiled Tibetans have encountered significant difficulties in their efforts to accommodate such a model of citizenship, including challenges faced in translating and interpreting the concept of 'the citizen' into both the Tibetan language and Tibetan cultural contexts (see Frechette 2006).

On the one hand, the discourse of citizenship is, like the system of justice commissions (Chapter 4) and the practice of democratic politics (Chapter 5), poorly developed given how recently the concept was adopted by the community and the legal restrictions faced in the situation of exile. On the other hand, the practical implications of Tibetan citizenship are widely understood and engaged with by members of the exile community, and offer a revealing insight into the construction of exile Tibetan political subjectivity. Tibetan citizenship is materialised in a pseudo-passport, the 'Green Book', and the annual payment of voluntary contributions (*chatrel*) to the TGiE. Alongside the establishment of citizenship, the Green Book was also instituted outside the homeland as no universal identity document existed in pre-1959 Tibet, and Tibetan passports were only introduced in 1947 and issued by the *Kashag* to just four Tibetan diplomats. The Green Book and *chatrel* systems were initiated by a group of exile Tibetans in 1972 under the auspices of the 'Tibetan Freedom Movement' with the rationale of encouraging Tibetans to express their loyalty to the TGiE by making a financial contribution to its running costs. In 1991, the holding of this document and contribution of *chatrel* became enshrined in the Charter as one of the main duties of Tibetan citizens in exile, and, in 2004, the TGiE's Department of Finance became the sole authority to issue Green Books. According to TGiE officials, more than 90% of exile Tibetans hold a Green Book, and these documents are issued to all Tibetan children born in exile and to each new arrival from Tibet. Meanwhile *chatrel* is, as members of the administration are keen to stress, a modest but symbolically important contribution to the TGiE's budget.[9] In light of assertions that systems of documenting individuals' identity are based on mechanisms of legibility that extend the reach of the modern state (Caplan & Torpey 2001; Scott 1998), this increasingly standardised administration of Green Books and *chatrel* can be seen as a key performance of stateness.

On one level it is easy to dismiss Tibetan citizenship and the Green Book as merely symbolic. Like other exile governments, the TGiE cannot legally defend or protect its own citizens nor, given its lack of jurisdiction over territory, guarantee them basic rights of abode. In addition, the rights and obligations of Tibetan citizenship are not legally enforceable: no Tibetan can travel on a Green Book as it is not recognised by any state, and *chatrel* is deemed 'voluntary' as there is no legal means of compelling payment. Yet, despite these sizeable limitations, holding a Green Book and paying *chatrel* do bring important material benefits, and exile Tibetan citizenship does have a number of parallels with conventional state citizenship. Interviewees consistently spoke of the Green Book as their 'Tibetan passport' and their duty to pay 'taxes' to their government, while the Department of Finance (2005) is careful to stress that *chatrel* is not a 'donation' as this connotes a different form of relationship, and crucially not one based on obligation and duty. Semantics aside, perhaps more importantly the TGiE exercises significant moral and social coercion regarding the implementation of Tibetan citizenship. The holding of this document and payment of this 'contribution' are essential to functioning in the exiled community, from gaining admission to Tibetan schools and accessing TGiE-run welfare services, to being eligible for TGiE-administered stipends and government jobs, and voting and standing in exile Tibetan elections. It is the privileges and responsibilities intrinsic within Tibetan citizenship that therefore means that the Green Book system is a key mechanism through which the TGiE attempts to secure loyalty and manage its exile population, confirming Sidhu and Christie's assertion that citizenship is 'an assemblage of techniques and technologies aimed at producing governable subjects' (2007: 15).

Central to this is the fact that the Green Book acts as both a unifying and exclusionary marker of Tibetan identity. For, although the TGiE cannot control residence in or prohibition from any territory, it can determine membership of its exile community. As well as excluding 'obvious' non-Tibetans, eligibility for Tibetan citizenship differentiates between Tibetans living in South Asia and ethnically similar groups such as Ladakhis, Bhutias and Sherpas, who are Indian or Nepali citizens. Members of these communities often follow Tibetan Buddhism and are eligible to join Tibetan monasteries, nunneries and schools, but, under the criteria for Tibetan citizenship, are generally not entitled to hold a Green Book. As such, in doing more than simply asserting 'Tibetan-ness as an ethnic identity within the pluralistic Indian and Nepali states in which they now live' (Shneiderman 2006: 16), Tibetan citizenship is both constituted through and, in turn, produces a series of 'other subjects from whom the subject of a claim is differentiated' (Isin 2008: 18).

On the flip side, Tibetan citizenship is also deliberately inclusionary, with every Tibetan citizen as defined in the Charter in theory being eligible for a Green Book, although in practice it is only those in exile who are able to enact the obligations and enjoy the rights of this citizenship. In contrast to refugee status and identity documents issued by the Indian state, which vary according to where an individual was born and/or when they came into exile (see McConnell 2013a), Tibetan citizenship and the Green Book are therefore accessible to all exile Tibetans, including 'newcomer' refugees. Indeed, when arriving in India Tibetans from Tibet need to 'become' Tibetan citizens before they can start to engage practically with the exile community. They might have entered the territory of and sought asylum in India, but they are also required to seek membership of the Tibetan 'state' in exile and to become 'bona fide' Tibetan citizens.

The 'nationalisation of citizenship' (Isin & Turner 2007: 11) in this case means that, to a large extent, the pseudo-*legal* identity regime of Tibetan citizenship has become synonymous with the *political* identity regime of Tibetan civic nationalism. As noted above, a central element of this discursive construction of Tibetan nationalism has been the TGiE's fostering of a single Tibetan identity, and universal citizenship has been a key way of trying to forge a sense of national unity across the diaspora. This endeavour resonates with Corrigan and Sayer's analysis of state activities 'regulating into silence identification based on difference and promoting integrative categories of official discourse—the citizen, the voter, the taxpayer, the consumer' (1985: 198), and a number of Tibetans I spoke to in the settlements perceived the Green Book as a material symbol of their national identity. Pride is taken in presenting it when required. In many homes Green Books are carefully wrapped and kept on a high shelf, with height above oneself denoting importance and reverence in the Tibetan Buddhist context. The valuing and treasuring of these identity documents has similarities with Gastón Gordillo's (2006) research on the relationship between the indigenous people of the Argentinean Chaco and their *documentos* in terms of how this community has 'internalized their past alienation from citizenship rights through the fetishization of those objects long denied to them: identity papers' (2006: 162). Whilst the Argentinean *documentos* were issued by a recognised state, the revered treatment of Green Books likewise represents a state-like fetishism of the institution of the TGiE (Navaro-Yashin 2005; Taussig 1997).

Indeed, by framing this identification regime in state discourses of citizenship, the TGiE thus attempts to assert the political right to classify persons as citizens. Central to this is the claiming of legitimacy. The holding of a Green Book and contributing *chatrel* are the primary mechanisms by which exile Tibetans recognise the TGiE as their legitimate government, thereby solidifying the link between the diaspora and the exile administration rather than other organisations or institutions. Meanwhile, the TGiE claims that 'from the legal point of view those who contribute towards *chatrel* and hold a Green Book are recognised as bona fide Tibetans in exile' (Department of Finance 2005: 2). Significantly, the Green Book as a signifier of 'authentic' Tibetan identity is not restricted to the exile community but is increasingly acknowledged and tacitly recognised by external actors, albeit in starkly different ways. With regard to China, given that the Chinese government publicly declares TGiE to be a 'splittist' organisation (Phuntsog 2005), Tibetans could be persecuted by the Chinese authorities if they return to Tibet and are found carrying a TGiE-issued Green Book (Home Office 2003: Section 6.304). Equally revealing, the Immigration and Refugee Board of Canada declares that 'one of the best ways to determine if a person is a bona fide Tibetan in exile is to see if they have a "Green Book"... the authenticity [of which] can be verified by the Office of Tibet that issued the document' (Immigration and Refugee Board of Canada 1998). Such acknowledgment of an 'official' Tibetan identity in exile – albeit not recognition of Tibetan citizenship per se – therefore challenges assumptions that stateless individuals are '"invisible" because they do not conform to the modern political imaginary' (Mandaville 1999: 663). The validation of Tibetan identity by the TGiE can thus be read as an important attempt to create an international political (if not legal) presence for this community.

The TGiE's desire and ability to shape political subjectivities is not, however, confined solely to the construction of Tibetan citizenship in exile. The identity category of 'Tibetan citizen' tells only half of the story, for these individuals also have a complex relationship with the category of 'refugee'. Given India's decision to sign neither the 1951 UN Convention on the Status of Refugees nor the 1967 Protocol – and thus its lack of refugee law – exile Tibetans legally come under the 1946 Foreigners Act. These individuals are thereby classified as 'foreign guests' in India, even if they have been born there.[10] Yet, while denied the *legal* status of 'refugee', the *political* label 'Tibetan refugee' is widely used both by Indian authorities, and across the Tibetan community. This active adoption of refugeehood – subject positions conventionally associated with vulnerability and dependence – has revealing resonances with Diane Taylor's analysis of the Argentine 'mothers of the disappeared' (*Los Madres*) who protested against the disappearing of the nation's young men during the 'dirty war' of the 1970s. Just as the women of *Los Madres* performed according to a script of 'Motherhood', which arguably obscured differences among women and 'limited the [Resistance's] arena of confrontation' (Taylor 1995: 300), Tibetans in India have, under the direction of the TGiE, performed under the collective scripts of refugeehood and statelessness.

The material manifestation of Tibetans' refugee status within India is the 'Indian Registration Certificate for Tibetans', or 'RC', which is issued through the local Superintendent of Police, needs to be renewed every 1 or 5 years, and permits the right to reside in the area where it is registered.[11] Trying to acquire an RC and struggling to function in India until one is issued is a source of anxiety and vulnerability for many recent refugees. However, at the same time, the annual renewal of this document was seen by a number of interviewees as a valuable reminder of their refugee status and the fact that India is not 'home', even if they were born there:

> Every year we Tibetans have to renew our RCs – a formality like this... and each time this reminds me I am a refugee and each time it provokes a strong emotion. ... The RC – this is the real identity of Tibetans in India – it reminds us that we have to go back to Tibet. That we are visitors in India and our stay cannot be permanent. (Shopkeeper, Dharamsala, February 2006)

Thus, while the Green Book proves that an individual is a 'bona fide Tibetan', the RC verifies de facto refugee status in India and is thus 'expressive of a cultural, ethnic and national identity, an allegiance to the past and a candid avowal of dedication to Tibet's future freedom' (de Voe 1987: 56). This situation is not unique to exile Tibetans. It has significant parallels with, for example, Liisa Malkki's research among Hutu refugees in western Tanzania, where she observed that refugee status becomes 'valued and protected as a sign of the ultimate temporariness of exile and of the refusal to... put down roots in a place to which one did not belong' (1992: 35).

The TGiE itself has actively promoted this correlation between the stateless status of Tibetans in India and loyalty to the broader Tibetan cause. Not only are valid RCs required from Tibetans in India in order to apply for TGiE-administered scholarships, government jobs and, indeed, to acquire a Green Book, but also the exiled leadership has urged Tibetans to retain this 'refugee' status even if they are eligible for Indian citizenship under the 1955 Citizenship Act (Shiromany 1998: 242, 263, 337).[12] Setting this dissuasion towards Indian citizenship in a broader context, Melvin Goldstein noted in the early years of exile that

> a consequence of this policy (whether intentional or not) is the greater dependence of the refugees on the Dalai Lama's Government. Since as individuals Tibetans are stateless 'guests'... their strength lies in their collectivity and is precisely the role of the Dalai Lama's Government to organise and represent them collectively. (1975: 24)

Or to put it another way, as long as Tibetans in India are registered as foreigners, their 'primary legal obligation as citizens can be seen as being towards TGiE and that this fact significantly bolsters the legitimacy of TGiE' (Tibetan Political Review 2013a). Through the scripting and promotion of refugeehood alongside, and indeed constitutive of, Tibetan citizenship, the case of the TGiE thereby challenges the dualism of state and stateless identities in particular ways. This is not the hybridisation of citizen and refugee identities – each is articulated in specifically conventional forms – but rather the co-scripting of them in ways that mutually reinforce these particular subject positions and enhance the role of the exile government as the legitimate representative of this population.

Conclusion: Believing in State-like Scripts

In examining the TGiE's governance of lives and livelihoods in exile, this chapter has examined how different domains in this exiled polity are constituted as governable. Though the TGiE faces considerable limitations in providing for and governing its population, welfare state and nascent economy in exile, nevertheless its partitioning out and attempts at managing these domains of governance are significant. Following Rose and Miller's assertion that '[K]nowledges of the economy, or of the nature of health, or of the problem of poverty are essential elements in programmes that seek to exercise legitimate and calculated power over them' (1992: 182), it is through such strategies of governmentality that the TGiE is legitimised as a governing authority. But to what extent is this exercise of authority *state*-like?

On the one hand, TGiE's stateness is frequently undercut by its inability to legally compel or coerce its citizens to obey its orders, from cooperating with the census to remaining in the exile settlements. As such, with law as the limiting factor in this case, the TGiE can enact the state-as-provider role, but not state-as-protector. However, on the other hand, the TGiE enacts a number of performances of statecraft and, though its modes of governmentality do not make this a state, the process of constructing a population in exile and scripting a universal ideal of citizenship based on individual freedoms and equal rights suggests that the TGiE is discursively positioning itself as state-*like*. Moreover, institutions and signifiers of statehood are strategic resources for this unrecognised polity. For, in addition to underpinning the practical project of governance in exile, the TGiE's enactments of stateness constitute an important role in seeking legitimacy and demonstrating its ability to govern competently. We might, therefore, think in terms of governmentalisation to practise and rehearse the state (McConnell 2012).

This case also offers a revealing lens on the relationship between governmentality and territory. In light of the exile government's lack of jurisdiction over territory, the role of the population in exile as a realm over which TGiE can govern is amplified in this case. Nevertheless, this does not prevent the exile administration from employing territorialising strategies of governance. As illustrated above, TGiE's governance is often most effective when it is territorialised within the spaces of the exile settlements. Moreover, through its collation of population data, production of IDPs, and drafting of school curricula, the TGiE headquarters in Dharamsala has become, to use Latour's term, a 'centre of calculation' (1987: 215). Thereby acting as a locus of government for this exiled population, and being able to govern at a distance through such practices, this crucially reinforces the state-like nature of this exile polity.

In thereby offering grounded examples of how the state as an 'effect' is constituted, this chapter has highlighted the importance of a relational and multi-scalar approach to understandings of the state. With the scripting of welfare, citizenship and refugeehood creating Tibetan political subjects in exile, it is through the calling into existence of a set of obligations and rights between the TGiE and its (quasi) citizens that this polity emerges as a state-like institution. Crucially, as illustrated by the sacrificing of material advantages of Indian citizenship in order to retain the 'refugee' label, these relationships are based on trust and belief in these scripts and their author – the TGiE – which in turn goes some way to legitimising this polity. Whilst this belief certainly comes with caveats – not all exiled Tibetans buy into the TGiE's vision and this is a rehearsal that is missing significant players (Tibetans in Tibet) – nevertheless the social contract established with its diaspora offers the TGiE an important mechanism for claiming political agency.

Endnotes

1   The Tibetan population continues to be monitored by the Indian authorities through the administration of Indian-issued 'Registration Certificates' and, in 2011, Tibetans who had resided in India for more than six months were included for the first time in the Indian National Census (*The Tibet Post* 2010).
2   Owing to the political unrest in Tibet in spring 2008, the second census was delayed until April 2009. Results were published by the Planning Commission in December 2010.
3   The 2004–2007 Integrated Development Plan outlined 534 projects and programmes with a total cost of 1.04 billion Indian rupees.
4   I am cautious about drawing parallels between *kiduks* and Chatterjee's (2004) analysis of local welfare associations in informal urban settlements being key intermediaries in 'political society' as, whilst these associations are based on ethnic/regional identities, they are not used to make collective claims vis-à-vis TGiE programmes.
5   Pseudonyms are used.
6   This is a process that assists NGOs' registration under the Indian government's Foreign Contribution (Regulation) Act 1976, which in turn permits them to solicit funds from abroad.
7   The full guidelines were published in the 8 August 2007 edition of *Bod-mi Rawang*, the TGiE's Tibetan-language weekly.

8   During the first decade of exile, $5,300,000 was received in direct aid from the US government (Grunfeld 1987: 189–190), and in the period 1995–2000 74.9% of the total funding for project expenditure and 30% of funding for recurrent development expenditure came from foreign donors (Finance *Kalon*, 10 April 2006).

9   *Chatrel* rates vary according to age, place of residence and income. For adult Tibetans resident in India, Nepal and Bhutan the rate is 58 Indian rupees per year, compared to US$96 for those resident in other states. For those on salaried income the rate is 4% of their basic salary. As of 2013 *chatrel* contributions constitute around 8% of the TGiE's annual revenue.

10 For a discussion of India's ad hoc and discriminatory treatment of different 'refugee' communities see Chimni (2003) and Samaddar (2003).

11 De facto refugee status (and the RC), generally apply only to those Tibetans who arrived in India between 1959 and 1979, and their children. The improvement in Sino-Indian relations in the 1980s and the increasing numbers of Tibetans coming into exile following the liberalisation of Chinese policy in 1979 meant that Tibetans arriving after this period have often not been recognised as refugees by the Indian government and, although they are generally allowed to remain in India, have no legal status (McConnell 2013a: 972).

12 There are signs that this stance on Indian citizenship may be beginning to change, especially in light of the material and economic benefits that Indian citizenship would bring to the Tibetan youth, and a ruling by the Delhi High Court in December 2010 that challenged the Ministry of External Affairs' denial of an Indian passport to Namgyal Dolkar Lhagyari, a Tibetan born in India in 1986 (Bodh 2011).

Chapter Seven
Audiences of Statecraft: Negotiating Hospitality and Performing Diplomacy

It is day three of a week-long visit of British MPs to Dharamsala. A packed schedule of meetings has been arranged with TGiE officials, as well as visits to Tibetan cultural institutions and 'interactions' with recent refugees from Tibet. But today is the most important engagement: a visit to the Tibetan Parliament-in-Exile, the MPs' hosts for this trip and their counterparts for an ongoing parliamentary exchange programme between Dharamsala and Westminster. The Speaker of the Tibetan Parliament, Penpa Tsering, ushers the British delegation into the chamber. It is laid out in a horseshoe of seats for the *Chitues*, with the speaker's chair at the top – under a photograph of the Dalai Lama and between two Tibetan flags – and benches at the back for the press. The British MPs, conspicuously addressing Penpa as 'Mr Speaker', ask about the Tibetan electoral system and how parliamentary business is conducted. The potted history of Tibetan democracy that is given is one regularly recited to visitors but, keen to make the British guests feel at home, the TPiE is pointedly described as the 'grandchild of Westminster', following the line of descent via the Indian *Lok Sabha*. We file out of the chamber and into the parliament's lobby where we're joined by a handful of *Chitues*, here to meet and greet their British opposite numbers. The conversations turn to electoral reform – in both parliaments – Scottish devolution, the stalling of Sino-Tibetan dialogue and recent self-immolations in Tibet.

Such formal performances of Tibetan statecraft have formed a core part of the exiled leadership's activities since establishing its base in India in 1959. Perhaps the most iconic images of the Dalai Lama are ones of him shaking hands with a 'who's who' list of political leaders, from George W. Bush to Nelson Mandela,

Rehearsing the State: The Political Practices of the Tibetan Government-in-Exile, First Edition. Fiona McConnell.
© 2016 John Wiley & Sons, Ltd. Published 2016 by John Wiley & Sons, Ltd.

Jawaharlal Nehru to Angela Merkel. With the handover of political authority from the Dalai Lama to the *Sikyong* and his *Kalons* in 2011, the elected Tibetan leadership has stepped up the administration's international outreach. As part of a strategy to bolster their credibility and legitimacy as the (new) faces of Tibetan politics, the *Sikyong*, speaker of the TPiE and International Relations *Kalon* have each embarked on lengthy tours of North America, Europe, Australia and East Asia, where they have met with as many political leaders, legislators and human rights advocates as they can. They are 'out of station' so much that some in Dharamsala joke that the TGiE has shares in Air India. Yet this attempt to strengthen the international profile of the TGiE and its leadership is happening at a time when negotiations with the state that matters the most for the future of Tibet – China – have ground to a halt. The last round of talks was in January 2010, and in May 2012 the two envoys of the Dalai Lama resigned their positions, frustrated with China's unwillingness to discuss anything pertaining to the future political status of Tibet.

The previous three chapters have each looked at different aspects of the crafting of this state-like Tibetan polity in exile: its territorialising strategies, establishment of a functioning bureaucracy and fostering of state-citizen relations. This chapter turns to a more intuitive reading of statecraft in terms of the discursive practices of state elites to construct polities within the international sphere (Ashley 1988; Campbell 1992; Doty 1996). In bringing centre-stage the TGiE's interaction with a range of external actors, this final take on rehearsing stateness thereby looks beyond the exile Tibetan community to focus on how this exiled government positions itself on regional and international stages: the strategies it employs, the challenges it faces and the forms of legitimacy it draws on. Three particular spheres of interaction – or 'audiences' – are crucial both to the TGiE's ability to operate in state-like ways and to the furtherance of its long-term goals: the host state India, the occupying state China, and the international community, broadly defined. At first glance discussing audiences in the context of 'rehearsal' seems somewhat misplaced. Rehearsal is conventionally undertaken away from and prior to the critical gaze of external spectators. But as this study has sought to argue, the TGiE is rehearsing stateness both as a training exercise for its diaspora to practise the skills of democratic governance, and to demonstrate its legitimacy as representatives of Tibetan people, its competence as a governing institution and its vision for an alternative form of governance in Tibet. In these latter functions rehearsal itself is deliberately performative. As such, engagement with – and the soliciting of approval and support from – audiences that are perceived to matter takes on a heightened importance. Following Jeffrey Alexander's (2011) observation that audiences of performances of politics can be attentive or uninterested, focused or distracted, sceptical or supportive, this chapter highlights just such this heterogeneity across the TGiE's audiences and traces how the exiled Tibetan leadership tailors not only the performance of its interactions, but also the image of the TGiE that it presents to each.

I turn first to the host state India, upon which the TGiE is dependent in order to exist in its current form. Focusing on discourses and performances of hospitality, this section traces the complex and inherently fragile relationship between the exile Tibetan and Indian administrations at a range of scales. Secondly, as with other aspirant governments seeking political and material support in order to gain or recapture political power, cultivating relations with recognised states and inter-governmental organisations forms a core part of the TGiE's remit and activities. This section examines the tactics that the TGiE employs to foster international support and the obstacles it encounters, focusing both on a range of diplomatic institutions and performances, and on the TGiE's strategic promotion of discourses of 'good governance'. Finally I turn to the omnipresent but elusive audience of China. Leaving this 'audience' until last is not to downplay its significance. Quite the opposite for, in tracing Sino-Tibetan contestations over the reincarnation of the Dalai Lama and the series of thus far unfruitful Sino-Tibetan talks, I raise the possibility that the very state-like-ness of the TGiE proves a significant barrier to engagement with the audience that matters most: the Chinese authorities. As such, the chapter concludes by reflecting on the implications of the TGiE's shifting exilic priorities, and visions for the future for the rehearsal project itself.

Being Exiled 'Guests' in a Host State

'[A]t least one host state is the *conditio sine qua non* for a government-in-exile' (Reisman 1991: 240) and, as Yossi Shain outlines in his discussion of political exiles and national loyalty, host states can adopt a range of different attitudes towards exiles operating from their territories, from using 'the exile groups as propaganda tools to repudiate opponent regimes ... [to] bargaining chips in international conflicts, or even as fighting forces in foreign battles' (1989: 119). The relationship between an exiled administration and the state within which it is based thus offers a revealing insight into the foreign policy priorities of the host, and is central to the functioning (or not) of exiled polities. As I shall sketch out here the relationship that the TGiE has with its host state India underpins questions of temporality – the host-guest relationship is never intended to be a permanent one – the politics and the ambiguity of recognition, and the extent to which the TGiE has autonomy to engage with state-like practices whilst operating within another sovereign state.

Discussed in detail later in this chapter, the TGiE's relationship with Western states and the 'international community' more generally has been one conducted primarily in public, and with the explicit aim of garnering support for the broader Tibetan cause. However, with an inherent power imbalance between the actors, and often complex and tense geopolitical contexts, the relationship between an exile administration and its host state is never going to be characterised by anything close to conventional diplomatic relations. Fundamentally not one of equals, the relationship between the TGiE and the Government of India is framed within official discourses on both sides as between an extraordinarily generous and tolerant host, and grateful and largely obedient guests. Indeed, in many ways the relationship between an exiled community and its host state is a quintessential example of hospitality, and this is a concept that opens up productive lines of enquiry for exploring the feasibility and the politics of stateness in exile. Hospitality has been the focus of considerable attention in recent years across the social sciences and humanities (e.g. Benhabib et al. 2006; Candea & da Col 2012; Rosello 2001).Geographers, often drawing on Jacques Derrida's accounts of hospitality, have been particularly interested in intersections between this concept and questions around state borders, the rights of belonging and refugee resettlement (see Darling 2010; Mountz 2011; Ramadan 2008). For Derrida there are two distinct aspects to hospitality: a '*politics* of hospitality... [which] involves limits and borders, calculations and the management of finite resources, finite numbers of people, national borders and state sovereignty' (1999: 12, my emphasis), and an *ethics* of hospitality, expressed as an outlook of openness and acceptance. Following Ruth Craggs' observation that hospitality is 'better understood as always moving and shifting between these poles' (2014: 90) of an instrumental practice and an ethico-political position, in what follows I sketch out the performances, spaces and temporalities of hospitality within Indo-Tibetan relations.

Various explanations are posited as to the foundations of this hospitality, from long-standing spiritual connections between Tibet and India, to the fact that Tibetans are largely seen as model refugees (Fürer-Haimendorf 1990) and India's so-called 'national tradition' of providing shelter (Subramanya 2004). As a retired official from the Indian Ministry of External Affairs put it:

> Indians in the central government and in individual states when they harboured the Tibetans they were motivated by these several factors, by the humanitarian cause, by some sense of obligation, and the fact that India is the natural land of their rehabilitation (September 2012).

Whilst hospitality is not a prerogative of any particular region or nation (Hertzfeld 2012), there is a prevailing Indian discourse of treating 'the guest as god', and the provision of hospitality to refugees has been posited as an integral part of Indian postcolonial politics (Samaddar 2003). However, this is not an ethics of *unconditional* hospitality. India has played host to many displaced communities since its independence in 1947 but, as noted in the previous chapter, as a non-signatory to the 1951 Convention on the Status of Refugees and the 1967 Protocol, India has no national legislation regarding refugees and is not bound by international norms. Therefore, although the Indian judiciary has to date overwhelmingly acted to protect 'refugees', this is essentially a regime of charity, not a regime of rights. As a result of the state having the power to decide to whom to extend or deny hospitality, this has led not only to the differential treatment of various displaced groups in terms of residency and employment rights but even to different classifications (see McConnell 2013a). With Tibetans in India being dealt with under the 1946 Foreigners Act, these individuals are 'guests' in a legal as well as figurative sense, and experience the suspension of social and political rights that comes with this status (see Pitt-Rivers 1968: 24). This includes restrictions on land and property ownership, ineligibility for government jobs and travel constraints (Tibet Justice Center 2012: 12).

Just as the ethics of hospitality are complex and ambiguous in India's relations with Tibetans, so is the instrumental nature of this hospitality. Dawa Norbu (2004) suggests that India's generosity can, in part, be attributed to a perceived need to compensate Tibetans for India's withholding of diplomatic and military support for Tibet during the turmoil of the late 1950s. *Lok Sabha* debates regarding Tibetan refugees in the late 1950s and early 1960s certainly illustrate this, but there are also indications of a more politically strategic reading of India's hospitality whereby the presence of the Dalai Lama, the TGiE and over 100,000 Tibetan refugees on Indian soil in effect acts as a bargaining chip in India's relations with the People's Republic of China. The 'Tibet question' is frequently referred to as a 'thorn in Sino-Indian relations' in New Delhi or, as one retired Indian government official put it, 'India is preserving the "Tibet card" for the future, and will only use it when it is in India's interest' (September 2011). This framing of the TGiE as a 'bargaining chip' plays into the familiar trope of governments-in-exile as pawns rather than players in geopolitics. Whilst, as demonstrated in the preceding chapters, the TGiE is certainly not devoid of agency, this conventional understanding of exiled administrations usefully highlights their insecurities and vulnerabilities.

With the future of the TGiE and its community in India being to a significant degree contingent on Sino-Indian relations, various hypothetical future scenarios are deliberated within Tibetan and Indian policy circles. A broad spectrum of such suppositional thinking can be charted. At one end are positive scenarios for the Tibetan community whereby Indian military and economic powers increase to parallel China, and the Indian government uses this confidence and authority to fully recognise the TGiE and even consider arming a Tibetan insurgency. At the other extreme, New Delhi bows to Beijing's diplomatic, military and economic pressure, shuts down the TGiE and expels the Tibetan refugees. According to senior Tibetan and Indian officials, India is currently navigating a classically Buddhist 'middle path' position vis-à-vis Tibet, reflecting that 'to date there has been a happy coincidence of interest between the Tibetan people and India' (former Indian diplomat, September 2012). However, this stance is far from fixed. Improved trade relations with China in recent decades may have somewhat sidelined the 'Tibet question' but issues pertaining to Tibet have made it back onto the national agenda in recent years due to China's upstream control of key water resources and increasing concerns around border security

(Chellaney 2011; Kondapalli & Mifune 2010). The latter is an issue that, with its promotion of an assertive Indian presence on the regional stage, the administration of Indian Prime Minister Narendra Modi has been particularly concerned with.

An additional threat to the Tibetan presence in India has evolved from the longevity of this exile and signs of host fatigue. Reflecting age-old concerns of guests overstaying their welcome, hospitality is 'a process under constant negotiation' (Derrida 1997; Ramadan 2008: 664). Like many examples of long-term exiles, Tibetans in India are 'permanent-temporary guests' (Ramadan 2008: 659) unless or until they take Indian citizenship, and this situation of limbo is a politically fragile one. As an Indian journalist explained:

> Tibetans cannot expect to be treated in the same way here over time. The Indian officials dealing with Tibetans to date have been those who remember the Tibetans first arriving as refugees when they had nothing and had to struggle to make a life in India. And these officials have always been generous and understanding to the Tibetans because of their past, but soon these officials will retire from office and the second generation of Indians won't have this institutional memory. They have no memory of Tibetans as 'beggars' – all they see is successful refugees who are good at business and have a lot of help from India. So they're not going to be as sympathetic and understanding (November 2007).

Echoing these sentiments many older Indian MPs, civil servants, journalists and lawyers I spoke to expressed considerable empathy with the 'Tibetan people and the Tibetan cause', tracing this back not only to personal admiration and respect for the Dalai Lama but also to the national shame brought about by the 1962 Sino-Indian border war. Having not lived through this violation of *Hindi-Chini bhai-bhai* relations ('Indians and Chinese are brothers'), younger generations of Indian politicians and bureaucrats predictably have a different perspective on Sino-Indian relations: one where hospitality is more likely to be conditional and subject to capricious change. The post-Nehru shifts in Indian politics towards an increasingly less pluralist approach to ethnic and religious minorities has also shaped attitudes to the Tibetan community, although connections forged around religious traditions including pilgrimage have often placated relations with Hindu nationalist politicians.

The discourses and practices of hospitality thus have important temporal dynamics, but as geographers have been careful to point out, they also have key spatial dimensions (e.g. Barnett 2005). In this case it is the spaces that the Indian government has granted to the TGiE rather than more conventional sites of performed conviviality or welcome that are most revealing of Indo-Tibetan relations. As discussed in Chapter 4, though Tibetan settlements are on Indian territory and come under Indian legal jurisdiction, the host state's non-assimilative ideology means that Tibetans are able to maintain their culture and way of life within these spaces, and use these sites to rehearse Tibetan modes of stateness. As Michael Reisman notes, a 'key question for all ... governments-in-exile is how much autonomy, under what conditions and for what purposes they will be allowed' (1991: 243) and, though cautious not to overstretch the metaphor, we can perhaps view India as the (reluctant) owner of this theatre for the rehearsing of statecraft. The Indian government provides the rehearsal spaces, but it also enacts a constraining role on what the TGiE can and cannot do. As such, this offers a new angle on the power relations inherent within reciprocity and the tensions between 'spontaneity and calculation, generosity and parasitism, friendship and enmity, improvisation and rule' (Candea & da Col 2012: S1) that lie at the heart of hospitality.

Given the complexities and, at times, contradictions, of host-guest relations, it is perhaps not surprising that we can also catch glimpses of where the host-guest relationship in this case is inverted. Perhaps most obvious is the example of exile Tibetan institutions opening their doors to Indian host communities. Indians can use Tibetan health facilities, Tibetan schools admit a quota of Indian students (often from impoverished backgrounds) and Tibetan religious institutions accommodate large numbers of monks and nuns from Indian Himalayan Buddhist communities. Echoing Adam Ramadan's discussion of Palestinian refugees in southern Lebanon providing shelter for thousands of Lebanese displaced during the war between Israel and Hizbullah in 2006, in such examples Tibetans become 'hosts to their own hosts' (Ramadan 2008: 658). But what is notable here is that, rather than the spontaneous generosity of individuals and families towards those in acute need, this is primarily a formalised extension of hospitality that is articulated by the TGiE and is couched in discourses of state provision.[1] This is not to say that the power relations between India and exile Tibet are being reversed through such extension of welfare and education provision, but rather to suggest that this ability to act as host is evidence of the TGiE's expression of a degree of sovereignty (Shryock 2012).

It is also, of course, an instrumental and politically strategic act, and this can be traced on two scales. First, with the revival in Buddhism in many Indian Himalayan regions that have strong cultural, linguistic and religious connections to Tibet, the TGiE has framed its engagement with these communities as partly justifying the continued Tibetan presence in this particular host state. Second, at a local level extending hospitality forms part of the TGiE's strategy for ameliorating relations with particular Indian host communities. Given the duration of the Tibetan refugee stay in India the occurrence of communal tensions has been surprisingly rare. Outside of large cities, interactions between the refugees and the host population are often limited to everyday economic transactions. Yet the creation of job opportunities for local Indians as labourers on Tibetan farms and workers in Tibetan cafés, restaurants and carpet factories has proved a not insignificant factor in improving community relations. This is complemented by granting access to Tibetan-run facilities:

We do take some Ladakhi students in our school. The local community in this area is a poor community with little services, and there have been some problems with the Tibetans because they saw them come as refugees and stay in tents and now they are more successful. So sometimes there is some tension and so we admit some of their children to help these relations. (Principal of TCV School, Sonamling, May 2007)

However, with Tibetan settlements often perceived as enclaves of visible affluence compared to neighbouring host communities there have been cases of inter-ethnic disputes. The most notorious incident occurred in Dharamsala in April 1994 when an Indian youth was fatally stabbed by a Tibetan refugee, sparking rioting in the town during which the TGiE headquarters and many Tibetan shops and homes were ransacked and looted. As K. Dhondup noted at the time this incident was a 'rude awakening for the Tibetans in Dharamsala: an awakening to a reality long suppressed or simply forgotten. Tibetans have come as refugees and are expected to live as refugees' (1994: 18). Despite being a distinctly traumatic event, this disturbance did have constructive outcomes in terms of the establishment of Indo-Tibetan Friendship Associations – committees of community leaders that foster inter-communal cooperation – in areas where Tibetans have settled. I witnessed the qualified success of this strategy in 2007 when, after a violent altercation between a Tibetan and an Indian auto-rickshaw driver in Dharamsala, which prompted Tibetan boycotts of Indian taxis and shops in the town, leaders from the two communities held formal meetings and organised public apologies from the parties involved. Such choreographed displays of reconciliation and unity thus illustrate the continuing sensitivities as to how Indo-Tibetan relations are represented to audiences from both communities.

Apparent from this incident is a close working relationship between Tibetan officials and their Indian counterparts, and a rosy picture of such relations was painted by most of the Tibetan and Indian officials I spoke to. Within the Tibetan community, considerable fuss is made about visits of Indian dignitaries to Dharamsala and the settlements, and the ensuing public functions routinely result in gushing praise and set pieces on brotherhood, gratitude and mutual appreciation on both sides.[2] Yet whilst efforts are almost universally made at presenting 'good relations' between host and guest, *how* the TGiE is viewed and its role narrated vary considerably according to the scale of interaction with the Indian state. It is at the local level that relations between Indian and Tibetan officials are often most on a par. When undertaking routine duties, local level Indian bureaucrats frequently regard TGiE officials as 'government representatives' and thus as their official counterparts. Evidence of the currency of these 'official' titles was apparent when, in acknowledging the significant limitations her government faces, a researcher in the TGiE's Department of Information and International Relations (DIIR) explained that, our government can't help or intervene for its citizens officially because we are in India, they can only do it unofficially as a Tibetan helping a Tibetan. But at the local level sometimes saying you are from the 'Tibetan Government', or especially from the 'Dalai Lama's government', this makes the local Indian officials listen and can help the Tibetan's case (March 2007).

Yet, such tacit recognition of the TGiE as a government through Indo-Tibetan interactions at a local level is simultaneously denied at a national level. In relations between New Delhi and Dharamsala this is, of course, far from a conventional state-state relationship. This is exemplified by the fact that Indian and Tibetan politicians are conspicuously cautious in the terminology they use to describe the functioning of the TGiE. Since the Dalai Lama's arrival in India, the host state has resolutely refused to recognise this institution as a government (TPPRC 2006). Reflecting on his time in Dharamsala in the 1960s a former Indian government liaison officer to the Dalai Lama explained how,

> from our point of view he [the Dalai Lama] wasn't running a government-in-exile, rather he was running an administration to help his people. Politically we leave it to the Dalai Lama to resolve their problems ... we haven't stood in their way. It's not our policy to prop up the Dalai Lama as a sovereign leader, we just support him and respect his wisdom (May 2007).

Not only do such sentiments raise the question of how India will react when the current Dalai Lama passes away, but Indian unease with a government operating on its territory has been fully acknowledged by the Tibetan leadership. Although the TGiE is referred to as a 'government' (*shung*) in common parlance in the Tibetan community, the exile leadership have, in recent decades, deliberately not used the language of government to describe the political functioning of their administration. Rather the 'Central Tibetan Administration' (CTA) is presented to official Indian audiences as first and foremost a welfare organisation rather than political agitator. Likewise, in order to avoid appearing 'too legal' in the eyes of the host the Justice Commission serves demands for compensation rather than 'fines', and cases are heard before 'observers' rather than a 'jury'. This semantic cautiousness is significant in two regards. Firstly, it illustrates the TGiE's downplaying of its status and operations in an attempt to foster a 'conducive atmosphere' for the now stalled negotiations with Beijing. Secondly it is evidence of how the TGiE 'tiptoes' around the Indian government in order to safeguard the continuing Tibetan presence in India. So, for example, while *chatrel* is not strictly 'voluntary' (see Chapter 6), the term is used because to call for mandatory taxes within India would have been illegal.

Yet this is not to say that the Indian government does not take the TGiE seriously. Far from it. Not only does the TGiE's Department of Security liaise closely with their Indian counterparts but, as I have discussed elsewhere (McConnell

2013a), the Indian government also relies upon the TGiE to verify individual Tibetans' identities, particularly in the processing of Indian-issued identity certificates.[3] As such, the exile government in effect takes on a watchdog role, with the Indian administration relying on its literal and metaphorical 'stamp' of approval to corroborate a Tibetan's identity. Whilst not an endorsement of Tibetan administration as a government per se, nor indeed legal recognition of Tibetan citizenship, this reliance does indicate that the Indian government has a significant degree of trust in this exile polity, viewing it as a partner in particular security issues and the ultimate guarantor for its diasporic population.

In sum, officials from both polities describe a relationship that is conducted both officially and unofficially and is simultaneously benevolent and restrictive. It is a relationship that, from the Indian perspective, entails officially prohibiting Tibetans' anti-Chinese protests in India and, at the same time, effectively allowing an extra-territorial government of Tibet to exist on Indian soil and to engage with foreign dignitaries there: a 'duplicity' that certainly riles those in Beijing. A careful balancing act has thus far been achieved. The Indian government has permitted the TGiE 'a certain latitude without legitimising it' (former Indian Army General, September 2012). Key to understanding this host-exiled guest relationship is thus the fact that the TGiE's state-like practices and authority *as* a government to its people is never openly declared by the TGiE nor explicitly acknowledged or officially sanctioned by the Indian authorities. Bringing together the ambiguous temporality of hospitality with these performances of tacit sovereign authority, the idea of rehearsal – of this being the *practising* of statecraft – provides a useful explanatory angle on this situation. This is not to say that the performance is not real. Rather the TGiE's distance from the disputed territory of Tibet, their temporary guest status and their self-declared duty to take responsibility for their diaspora helps to ameliorate what is on paper a precarious existence within the state of India. But what of other audiences that the TGiE engages with out of choice rather than necessity? As I explore in the following section, in seeking to persuade key players in the international community of the legitimacy of the Tibetan cause a notably different register of rehearsal is employed.

Performing Diplomacy and 'Good Governance' for Western Audiences

> We Tibetans have considerable contacts with various foreign countries... These relationships firstly bring the Tibetans as a whole into contact and help the overall Tibetan cause. Secondly, they help to project that the Tibetans-in-exile have a parliament which has the qualifications of a democratic institution. The promotion of the parliament is not merely for the sake of name... we should deliberate upon ways of further strengthening these outside contacts. (Dalai Lama's speech at the Assembly of Tibetan People's Deputies, Dharamsala, 28 July 1993, reproduced in Shiromany 1998: 296)

It perhaps goes without saying that 'exile organisations ascribe vital importance to foreign support for their struggle' (Shain 1989: 110), and the TGiE is no exception. As the DIIR *Kalon* succinctly put it, 'our diplomatic corps are vital. They are our face and voice overseas, and are key to promoting our cause' (October 2012). Negotiating contacts with foreign – and in particular Western – governments has been a core activity since the early days of exile.[4] The challenge has been to keep these Western audiences interested and engaged, and to attempt to convert sympathetic platitudes into political and material support.

At first glance, the success of Tibetan engagement with the international community appears somewhat contradictory. On the one hand it has thus far failed to achieve any positive resolution to the 'Tibet issue'. Since passing resolutions in 1959, 1961 and 1965, the United Nations has been deafeningly silent on the issue of Tibet, a situation that is likely to continue given China's seat on the Security Council. Meanwhile, though many governments have passed resolutions condemning human rights abuses in Tibet, any support has been cautious and largely insubstantial for fear of angering Beijing and harming trade relations with China (Fuchs & Klann 2010). Yet, on the other hand, the exiled Tibetan leadership has been exceptionally successful at *sustaining* their cause in the eyes of Western audiences over past decades, and notably more successful than 'other exiles or minorities [e.g. Uyghurs or Mongolians] in denying legitimacy to China's rule over their homeland' (Tenzin Dorjee 2013: 65).

As a number of Tibet scholars have noted, the internationalisation of the Tibet issue was a strategic decision taken by Dharamsala in 1986–87 (Barnett 2001; Shakya 1999). This constituted a marked shift from an inward-looking focus on institution building and cultural preservation and external engagements that did not go much beyond the promotion of Tibetan Buddhism, to an explicitly political engagement with Western powers that were believed to be able to offer leverage with Beijing. The charismatic leadership of the Dalai Lama, the iconic 'national brand' of Tibet and international news coverage of ongoing protests in Tibet have been key to this. But so too has been considerable 'behind the scenes' work in fostering a network of sympathetic friends in the West, from political and religious leaders such as Václav Havel and Archbishop Desmond Tutu to figures in popular culture and members of foreign publics. In seeking to challenge the nature and the legitimacy of Chinese rule in Tibet – a strategy common across a range of exiled political groups (e.g. Watts 2004, on Kurdish transnational activism) – the TGiE and a plethora of Tibetan NGOs and Tibet Support Groups devote considerable time and resources to documenting what is happening inside Tibet, communicating this through public awareness campaigns and lobbying governments, the UN and multinational companies investing in Tibet (see Davies 2012; tibetnetwork.org).

Running parallel to this political activism is the exiled leadership's enactment of formal, state-like diplomacy. It is these diplomatic practices that direct our attention to the question of *how* the TGiE seeks to portray itself to potential sup-porters, and to the mechanisms through which the promotion of political legiti-macy is contingent upon demonstrating a competency in statecraft. With diplo-macy perceived to have a strong performative aspect (Shimazu 2014; Sidaway 2002) attending to this mode of political engagement thus offers an important lens through which to examine relations between statecraft and stagecraft.[5]

Diplomacy plays a central role in the conventional conferral of state legitimacy and functioning of the inter-state system. Taking the European state system as the traditional point of departure (Neumann 2012), scholarship on diplomacy has traditionally assumed the state to be a natural and bounded container for political activity. If we are to put this conventional understanding of diplomacy alongside the TGiE's lack of international legal status, the logical conclusion is that this polity's engagement with formal diplomacy is simply not on the cards. However, what is distinctive and revealing about the Tibetan case – and what sets it apart as a state-like institution rather than an NGO – are its *attempts* to maintain an international persona through a parallel engagement with and mimicry of official diplomatic practices, institutions and protocols (see McConnell et al. 2012). Of course, this foray of an exiled administration into the world of official state diplo-macy is far from unique. The governments-in-exile of occupied European states during World War II were primarily concerned with forging diplomatic connec-tions and securing support from their hosts and other allies. Likewise increasing academic attention has also turned to the paradiplomatic activities of sub-national and regional governments (Aldecoa & Keating 1999; Lecours 2002), not to men-tion the long historical engagement of indigenous communities with diplomacy (Beier 2010; Foley et al. 2013).

Forging diplomatic relations was considered to be of utmost importance to the TGiE in the early years of exile as, without coercive power and military capa-bilities – alongside a Buddhist commitment to non-violence – this administra-tion turned exclusively to leverages of 'soft power' (Nye 2004). Exile interna-tional affairs are handled by the 'Department for Information and International Relations' (DIIR), the largest ministry within the TGiE. Occupying a prominent position within the TGiE complex in Dharamsala, this department has its origins in the *chisee khang* (Foreign Relations Office), which was re-established in exile in 1959. With growing demand from the international community for informa-tion on Tibet in the 1970s and 1980s, the *chisee khang* was amalgamated with the 'Information and Publicity Office', which had been established in 1971. The DIIR's role reads, at first glance, like a checklist of conventional public diplomacy activities. These include liaising with international media, maintaining the exile government's website (www.tibet.net) and providing information on the human rights, political and environmental conditions in Tibet through a series of pub-lications. But the DIIR also engages in more traditional and formal modes of diplomacy.

At 'home', the Department serves as a protocol office of the TGiE, extending official reception to foreign diplomats, MPs and journalists visiting Dharamsala. Meanwhile, 'abroad' the DIIR maintains the TGiE's unofficial foreign missions, known as 'Offices of Tibet', in 11 cities across all continents. Although their diplomatic activities are restricted due to the TGiE's lack of legal recognition, these Offices nonetheless maintain direct contacts with governments, parliamentarians and NGOs, coordinate the Dalai Lama's official visits and act as a channel of news from Tibet. In addition to these embassy-like functions, and in light of limited funds, each Office was established with a clear goal in mind. For example, as part of an early effort to internationalise the Tibetan issue an Office was established in New York to lobby the UN. It subsequently served as the focal point for the increasing number of Tibetans settled in North America and, in 2014, was moved to Washington DC in order to focus on engaging with the US government. Similarly, the Office of Tibet in Brussels liaises with sympathetic parliamentarians in the European Union, and the Office in Geneva lobbies the UN Human Rights Commission.

Each Office of Tibet is led by a 'Representative of the Dalai Lama'.[6] Despite not having any status under the 1961 Vienna Convention on Diplomatic Relations, these members of the Tibetan diplomatic corps nevertheless describe receiving tacit recognition from host foreign ministries as being representatives of the TGiE. Their role also extends to representing Tibetans living in the states coming under the 'jurisdiction' of their posting. So, for example, the Office of Tibet in Tokyo seeks to establish bilateral relations with and is 'responsible' for Tibetans in Japan, South Korea, Mongolia and the Philippines, while the Office in London oversees relations with the United Kingdom, Ireland, Poland and the Scandinavian and Baltic states. With the growing Tibetan diaspora beyond South Asia, the consular-style assistance provided by these official Tibetan agencies is becoming increasingly important. Adhering to detailed directives from Dharamsala, such services include issuing and renewing Green Books, collecting *chatrel* and issuing 'bona fide Tibetan refugee' certificates required for some asylum claims. Thus, rather than sites of overt resistance or radical protest (cf. the Aboriginal Tent Embassy, see Foley et al. 2013), Offices of Tibet constitute important governmental spaces 'abroad'. They are spaces of representation, of legitimation, of lobbying and, importantly, of stateness, acting as polling stations during TPiE elections, and hosting community meetings and official functions. They also feel familiar as 'TGiE spaces', from the letter-headed paper to the photographs of the Dalai Lama and maps of Tibet on the wall, and the piles of government pamphlets.

But care must be taken not to paint a picture of these Offices as fully functioning embassies. A lot of doors remain shut to Tibetan diplomats, and forging relationships with officials from foreign governments is considerably more challenging than working with already sympathetic parliamentarians. For example, the Representative of the Bureau in Delhi noted candidly that his description of his relationship with, say, the British High Commission in India would depend on 'who was asking'. Officially there cannot be bilateral engagements given the British government's refusal to recognise the Tibetan administration *as* a government. However, as with the TGiE's relationship with the Indian state, unofficially there is often a close working relationship. This de facto recognition as diplomatic representatives and the ability to have a level of engagement with foreign governments is, as a former Representative at the Office of Tibet in Geneva outlined, based both on the 'good will' afforded by the international respect accorded to the Dalai Lama, and the fostering of personal connections over years if not decades:

> these days our representatives have fairly good access with most governments – we encounter a lot of interest with almost every government and also sympathy. Most representatives have no difficulties in getting appointments with government officials… For me personally I've been living and working in Europe for many, many years so I know many of the government officials in different places personally and so now I have this personal rapport which is so important. With this you can get access and trust (January 2012).

Although the fostering of personal connections is key to all diplomacy, it is a strategy that is of heightened importance in the case of 'unofficial' diplomats. This is especially so given that almost no one in the Tibetan diplomatic corps has had formal training in international diplomacy. Rather, over time and through experience, Tibetan diplomats have learned the language of international human rights and diplomacy. They have learned how to communicate with high-level officials and politicians, how to dress and carry themselves, and how to communicate their cause in ways that the international community can identify with. Therefore, while the financial, legal and political limitations of these Offices of Tibet are considerable, their existence and mimicry of formal state-like modes of diplomacy is nevertheless significant.

In addition to the network of Offices of Tibet, the TGiE employs a number of other strategies in order to establish diplomatic relations. A key audience in this regard has been legislators from states broadly sympathetic to the Tibetan cause, and the TGiE has organised a number of parliamentary exchanges in recent years. As sketched out at the start of this chapter, these involve delegations of parliamentarians visiting Dharamsala, and reciprocal visits of groups of Tibetan MPs to parliaments in the likes of the Netherlands, United Kingdom and Japan. These exchanges are promoted by the TGiE as opportunities for Tibetan MPs to learn about parliamentary best practice, and for foreign parliamentarians to hear first-hand about the challenges facing Tibetans in Tibet, and the achievements of the exile community. From my observations of such exchanges both in Dharamsala and Westminster, TGiE officials engage in two distinct, and seemingly contradictory sets of performances. One is that of conspicuous equality: they are meeting 'counterparts' who have also been democratically elected, represent constituencies and are professional legislators. The second is a presentation of themselves as still learning the art of democracy; of not being quite up to 'standard' yet and therefore deferring to their 'political elders' for advice and expertise.

Alongside these bilateral exchanges, the TGiE has also organised a series of World Parliamentarians' Conventions on Tibet, public performances of statecraft with the objective of garnering coordinated international support for Tibet. The first of these conventions was held in New Delhi in 1994, with subsequent meetings in Vilnius, Washington DC, Edinburgh, Rome and Ottawa, with over 100 delegates from 30 parliaments attending the latter. I was an observer at the fourth of these Conventions and, despite strong opposition from the Chinese consulate in Edinburgh and pressure from Chinese embassies on those states with parliamentarians attending, the event was an elaborate and carefully stage-managed affair. The hall in a leading Edinburgh hotel was decked out with the national flags of the MPs present – with the snow-lion emblazoned Tibetan flag centre stage – and, again, all parliamentarians were conspicuously treated as equals. As Yossi Shain notes, 'international forums have become the mecca for many exile groups that face difficulties in generating governmental support and thus search for alternative routes to the global audience' (Shain 1989: 128), and at carefully choreographed events such as this there is a palpable sense of Tibetan officials trying out the arts of statecraft. Within such spaces of diplomatic engagement, the TGiE staff work hard to foster belief both in their own exile administration, and in their vision for the future of Tibet. They are conspicuously playing their roles and promoting their script to an audience they hope can help turn their rehearsal into a reality.

Though sharing a number of characteristics of paradiplomacy, unlike regional governments the ability to engage in diplomatic practices is not devolved to the TGiE by a central state but is a strategy it enacts of its own volition. As such, these outward-facing performances not only imply some measure of international personality but also elicit tacit recognition. Thus, in seeking to enhance its own political effectiveness, the TGiE makes a concerted effort to mimic official diplomatic practices and institutions: key credentials of statehood that are perceived as essential modes of being in the interstate community. In using the term mimic I in no way want to imply a sense of parodying. This is an exercise in being seen to be *serious* about engaging in diplomacy in a state-like way in order elicit attention from international actors who are perceived to 'matter' (McConnell et al. 2012). Applying Bhabha's notion of mimicry – 'the desire for a reformed, recognizable Other, as *a subject of a difference that is almost the same, but not quite*' (Bhabha1984: 126, emphasis in original) – to diplomatic discourse, what we see here is the simultaneous promotion of official state diplomacy as the 'gold standard' to aspire to, and the unsettling of this 'ideal' by reducing the gap between the 'real' and the 'mimic'.

In addition to stage-managed protocols and the performative aspects of diplomacy the TGiE also seeks to persuade international audiences of its legitimacy through the promotion of narratives of good governance and universal rights. As Toni Huber has argued, the exiled elite have, since the 1980s, made a set of 'claims about the fundamental identity of Tibetans and the character of their traditional society and culture' (2001: 357), which include Tibetans being essentially peaceful people, Tibetan women enjoying more equality than in other Asian communities and environmentalism being an intrinsic part of Tibetan culture. Such claims are revealing in two ways. First, the scripting and promoting of these narratives by the exiled elite means that expressions of Tibetan identity and political history that contradict these dominant narratives are downplayed or even erased. This includes a series of 'public secrets' (Taussig 1997) around the violence of the Tibetan resistance movement and the continued presence of Tibetans in the Indian military (McGranahan 2010; Wangdu 2013) and an eschewal of discussing shortcomings in pre-1959 Tibetan society (Stoddard 1994).

Second, this is a classic case of telling particular (Western) audiences what they want to hear. Though arguably having 'little or nothing to do with so-called tradition and its continuity in the post-diaspora period' (Huber 2001: 357), these representations of Tibetan culture and identity are strategic attempts to align the Tibetan case to the standards – and thus expectations – of contemporary international politics (Sperling 2001). Since the second half of the twentieth century these have largely been premised on the twin issues of human rights and democracy and, in recent years, have coalesced around an increasingly agreed set of principles of 'good governance'. Used most frequently in the context of international development policy, and reflecting a Western bias within this domain, the qualities of good governance include participatory politics, transparency and accountability, the separation of powers, rule of law, environmentalism and gender equality. In recent years the TGiE has been in engaged in an almost tick-box exercise of meeting the criteria of these norms of good governance. For example, a quick scan down the list of 'development principles' outlined in the second Integrated Development Plan (IDP) – 'community awareness and participation', 'sustainability', and 'economic diversification' (Planning Council 1994: Section 2.2.2) – reads like a best practice guide for international development, albeit with 'added extras' specific to this case: 'preservation ... of Tibetan religion, culture ... and national identity' and 'return to a Free Tibet' (*ibid*). Beyond the IDPs the TGiE's recent promotion of professionalisation within the exile bureaucracy and fostering of participatory politics through devolution of authority to the settlements resonate closely with international expectations of emerging democracies.

Indeed, with the Tibetan elite having thus 'been learning the language of international politics as dominated by the West' (Anand 2000: 281) it is democracy – 'the leading standard of political legitimacy in the current era' (Held 2006: x) – that has been key. Alongside providing a mechanism for preventing a power vacuum after the death of the Dalai Lama (see Chapter 5), democracy in exile has also been initiated in relation to Chinese and international audiences. In terms of the latter, TGiE officials frame their community's transition to democracy both in terms of its 'uniquely' Tibetan qualities and its similarities to Western liberal democracies. Such promotion of democracy can be understood in terms of Appadurai's concept of 'ideoscapes':

> concatenations of images ... [that] are often directly political and frequently have to do with the ideologies of states and ... are composed of elements of the Enlightenment worldview, which consists of a chain of ideas, terms, and images, including *freedom, welfare, rights, sovereignty, representation* and the master term *democracy*. (1996: 36, emphasis in original)

Albeit careful not to appear *too* state-like to the Indian administration, it is this seeking of legitimacy – of the exile administration wanting to prove itself trustworthy and deserving of support in the eyes of international audiences – that has been the rationale for much of the TGiE's investment in state-like institutions, practices and discourses. With regards to China, since its inception Tibetan democracy has had the implicit aim of setting contemporary exile Tibetan political culture apart from both pre-1959 Tibetan politics and the modern Chinese state. A 'potent weapon for the cause and ... an absolute necessity for the credibility of the freedom struggle' (Norbu 2007: 35), Tibetan democracy is thus narrated as establishing a clear distinction between Dharamsala's progressive approach to participative politics and Beijing's one-party communist system.

Engaging the Occupying State: the Challenge of China

The Chinese government is an audience that is both ever present in exile Tibetan debates, and yet is a frustratingly elusive one when it comes to direct contacts. What I want to do in this final section is to sketch out the TGiE's (lack of) engagement with this key 'audience' and highlight important contradictions in the exile polity's performance of statehood vis-à-vis its occupying state. Two qualifiers are, however, worth noting. First, what follows discusses this relationship from the exile Tibetan perspective and therefore presents not only a one-sided account but also a somewhat typecast view of the Chinese state. For accounts that analyse both Tibetan and Chinese standpoints, those by Gray Tuttle (2005), John Powers (2004) and Elliot Sperling (2004) are particularly instructive.

Second, my focus here is specifically on the attempted engagement that the TGiE and the Dalai Lama have had with the Chinese government at various levels. It is important to acknowledge that this is far from the only contact that the exiled Tibetan side has sought with China over the decades. Rather, strategies of resistance and engagement range from the armed Tibetan resistance in the 1960s and early 1970s to Tibetan activists from the diaspora engaging in non-violent direct action during the Beijing 2008 Olympics and Tibetan exiles setting up small-scale educational and development projects in Tibet (e.g. www.machik.org; www.tibet-foundation.org). The Tibetan diaspora has also reached out to other exiled ethnic groups from the territory of the People's Republic of China (PRC). This has included co-founding the 'Allied Committee of the Peoples of Tibet, East Turkestan and Inner Mongolia' in Zurich in 1985 (expanded to include Manchuria in 1988), which, though no longer active, for several years produced a biannual journal – *Common Voice* – and issued statements of solidarity. Key figures from the exiled Tibetan and Uighur communities were also founders of the Unrepresented Nations and Peoples Organisation (UNPO) in The Hague in 1991, and these diasporas, sometimes alongside groups from Inner/Southern Mongolia, Taiwan and Chinese dissidents and democracy campaigners, co-organise conferences and hold joint demonstrations to protest visits of Chinese leaders to Western capitals.[7] However, the exiled Tibetan leadership rarely translates these public performances of solidarity into expressions of equivalence between other struggles for self-determination within China or into any form of joint lobbying vis-à-vis Beijing. When pushed on this, the TGiE officials I spoke to expressed caution about both 'diluting' their message about the exceptionality of Tibetan history, culture and religion and tainting the Tibetan case with direct associations with groups which have had more violent recent histories.

At first glance the relationship between the exiled Tibetan leadership and Beijing is a classic David and Goliath scenario. The former is overwhelmingly outmanoeuvred by the latter's economic, military and political clout both on the international stage and domestically. With regards to the Chinese government's views on the exile Tibetan polity, at face value these are boldly straightforward. Beijing publicly declares the TGiE to be a 'splittist' organisation (Phuntsog 2005; Wei 1989) and 'puppet government' of the Dalai Lama (interview with Zhu Weiqun in *China's Tibet*; Aiming 2011) and has no formal relations with it. Meanwhile the Dalai Lama is seen as a threat to the integrity of the Chinese nation and, consequently, meetings of foreign officials with the Dalai Lama are a constant source of bilateral diplomatic tensions with China. As a result, those international leaders who do agree to meet His Holiness do so increasingly in private and on the pretext of him as a religious rather than political leader.[8] Indeed, perversely, the fact that the Chinese authorities are so vehement in their criticism of the TGiE and its leadership and so keen to keep them politically isolated offers a clear illustration of the extent to which the exile government and the wider Tibetan freedom movement matters to Beijing.

It is the figure and institution of the Dalai Lama that remains of particular importance to China's stance both on exiled Tibet and on the future of the territory's leadership. In recent years, the question of Buddhist reincarnation has dominated 'exchanges' – mediated through formal statements and press releases – between the two sides. Since the fourteenth century, all lineages of Tibetan Buddhism have used reincarnation as the method of succession for high lamas.

Given the centrality of reincarnation to Tibetan leadership it has long been a political as well as religious practice (Goldstein 1989). As such, questions around what will happen when the current Dalai Lama passes away, whether and where a reincarnation will be found, and who holds the legitimate authority for recognising him/her form a key platform of contestation between China and the exiled Tibetan leadership.

As Ronald Schwartz argues, China has, since 1959, 'attempted to take over the role of legitimate patron of religion', thereby seeking to 'intervene directly in religious matters in order to shape Tibetan Buddhism to suit its political requirements' (1999: 237, 245). To date, the most overt intervention of the Chinese state into Tibetan Buddhist practices has been the dispute over the reincarnation of the 10th Panchen Lama.[9] The gulf created between the Chinese and exile Tibetan authorities over the former's appointment of their own Panchen Lama was further reinforced by Beijing's issuing of 'State Order No. 5: Management Measures for the Reincarnation of Living Buddhas in Tibetan Buddhism' in 2007. The order declares that only the Chinese government can recognise the reincarnation of a lama, including the Dalai Lama, and that all such individuals must be reborn within China. Order No. 5 was formally repudiated by exiled Tibetan Buddhist leaders and, on 24 September 2011, the Dalai Lama issued a further rebuttal that set out key practical steps regarding the future of Tibetan Buddhism. The declaration clearly spells out that only the Dalai Lama and, in his absence the Ganden Phodrang Trust, will have 'sole legitimate authority' for managing the Dalai Lama's lineage and the succession process. Further, His Holiness both explicitly excludes the PRC from intervening in the succession of the 14th Dalai Lama and uses his authority to delineate the future course of Tibetan Buddhism and, in turn, the Tibetan nation (see McConnell 2013b).

Reinforcing the assertion that legitimate authority for overseeing the process is placed firmly in Tibetan hands, the statement also indicates that, for the first time in six centuries, the Dalai Lama's successor will probably be identified before the current Dalai Lama passes away, will likely be an adult rather than a child, and will be identified outside of Tibet. China was quick to respond to the Dalai Lama's statement, with the Foreign Ministry spokesperson Hong Lei claiming that '[O]ut of ulterior political motives, the 14th Dalai Lama wilfully [sic] distorts and denies history... The title of the Dalai Lama is illegal if not conferred by the Central Government' (Lei 2011). As with the Panchen Lama, China is therefore likely to appoint its own successor, which raises the prospect of two Tibetan spiritual leaders, one recognised by Beijing, the other by Dharamsala. On a strategic level this is clearly an attempt by the Chinese government to reinforce its authority in Tibet, control the future Tibetan leadership, and split Tibetan loyalties (McConnell 2013b).

The figure and institution of the Dalai Lama also dominates what direct engagement there has been between the exile Tibetan community and the Chinese government. There have been two periods of 'delegation diplomacy' between the two sides since the late 1950s. Initial contacts were established in 1979 after the period of liberalisation following the death of Mao Zedong and the end of the Cultural Revolution. Gyalo Thondup, the Dalai Lama's elder brother, was invited to visit Beijing where he met Chinese leaders, and two further Tibetan delegations visited Beijing for exploratory talks in 1982 and 1984. Four fact-finding delegations were also sent by the Dalai Lama in 1979–1985 to ascertain conditions in different parts of Tibet. However, protests in Tibet in 1987 and Chinese crackdowns and the imposition of martial law that followed led to the suspension of relations.

In September 1987, the Dalai Lama presented his 'Five Point Peace Plan for Tibet' to the US Congressional Human Rights Caucus and, nine months later, announced his 'Framework for Sino-Tibetan Negotiations' in an address to the European Parliament in Strasbourg, whereby he formally renounced his government's previous demands for independence and launched his 'Middle Way Approach' (*Umaylam*). The so-called 'Strasbourg Proposal' called for the establishment of a 'self-governing democratic political entity... in association with the People's Republic of China', which would comprise all Tibetan-inhabited areas and whose government would have 'the right to decide on all affairs relating to Tibet and Tibetans' (Dalai Lama 15 June 1988). Premised on the Buddhist principle of seeking a path of moderation and conciliation rather than confrontation, the transposition of the 'Middle Way' into politics therefore represents a compact in which China would accede to 'genuine Tibetan autonomy' without compromising China's sovereignty, territorial integrity or security (DIIR 2005). In both meeting international expectations of good governance and positing a compromise position, the Middle Way approach has been actively promoted to foreign legislators and generally supported by the international community.[10]

However, the unilateral concessions made by the exiled leadership in both the Five Point Peace Plan and the Strasbourg Proposal were rejected by the Chinese leadership in 1990. Direct contact between the two sides did not resume again until 2002 when, at Beijing's invitation, a Tibetan delegation returned to the Chinese capital. According to one of the Tibetan envoys the aim was to 'create a conducive atmosphere for direct face-to-face meetings on a regular basis' and 'explain the Middle Way Approach' (Lodi Gyari, 28 September 2002), but more cynical accounts note that China was, at that time, keen to be seen to reach out to the Tibetans in light of international pressure in the run up to the 2008 Beijing Olympics (Smith 2008).[11] Since then eight rounds of talks have taken place between the United Front Work Department of the Chinese Communist Party (CCP) and representatives of the Dalai Lama, albeit with no preconditions set by the Tibetan side, nor any concessions made by the Chinese. Reflecting growing confidence in the skills of statecraft within the TGiE, Tashi Rabgey and Tseten Wangchuk note that 'in contrast to the highly personal nature of its earlier contacts with Beijing' this later phase of engagement has been 'considerably more systematic and professionalized' (Rabgey & Wangchuk 2004: 34).

However, the talks ground to a halt in 2010 after China yet again rejected the Middle Way Approach, stating that it would not discuss Tibet's current or future political status (only the conditions for the return of the Dalai Lama to Tibet) and would never agree to consider Tibetan areas outside the Tibetan Autonomous Region (TAR) as part of Tibet. A further blow to direct engagement came in May 2012 when the two envoys of the Dalai Lama – Kelsang Gyaltsen and Lodi Gyari –resigned citing 'utter frustration over the lack of positive response from the Chinese side' (*Kashag* press release, 3 June 2012).

The suspension of dialogue with China has not, however, prompted a change in policy on the exiled Tibetan leadership's side. The Middle Way Approach has, since the late 1990s, become the TGiE's sole policy regarding the future of Tibet.[12] As one Tibetan journalist put it, the Middle Way has increasingly come to be seen as '*the* national interest rather than one of several possible policy options' (September 2011). The dominance of this script is ensured by the leadership insisting that all TGiE staff promote it – whether or not they personally support it – and the exile government has launched a series of public awareness campaigns, producing pamphlets, short information films and a new website, and touring the settlements to explain the rationale of the policy.

In recent years, however, the Middle Way Approach has undergone significant and controversial modifications. In 2008 the Dalai Lama's position vis-à-vis the Middle Way Approach was clarified in a 'Memorandum on Genuine Autonomy for the Tibetan People', presented by the Tibetan envoys to their Chinese interlocutors as a basis for negotiation during the eighth round of talks.[13] Significantly, nowhere in this document is the word 'democracy' used to describe the governance of Tibetan regions within the PRC. Thus, at the same time as the Tibetan leadership is promoting its democratic practices of good governance in the present it is now seemingly denying such governance to Tibetans in a future homeland. Further 'watering down' of the Middle Way Approach came in May 2013 when the *Sikyong* Lobsang Sangay outlined his interpretation of the policy in a speech to the US Council on Foreign Relations.[14] His startling concessions of a time-limited period of autonomy, consenting to Chinese military deployment in Tibet and abandoning democracy in favour of the status quo of CCP governance structures provoked reactions of surprise, confusion and distress from a number of commentators within the community and prompted the issuing of a brief 'clarification'.[15]

On one level the Middle Way is promoted by the TGiE as a pragmatic political strategy to engage China in dialogue over the future of Tibet. With its small population size, lack of military force (and, indeed, commitment to non-violence) and an adversary as vast and internationally dominant as China, Tibetans arguably lack other feasible options. However, arguably more significant in explaining the diaspora's embracing of a policy that, in 25 years, has produced no tangible progress, is the fact it was proposed and continues to be endorsed by the Dalai Lama. With social prohibitions against criticising the Dalai Lama – and by extension his policies and his government – stronger than ever there is thus considerable social pressure on Tibetan individuals and organisations to publicly support the Middle Way Approach. This stifling of dissent has led to calls from some pro-*rangzen* (independence) supporters that freedom of expression is being undermined, and the reputation of Tibetan democracy threatened (*Tibetan Political Review* 2013b). Whilst this may be overstating the case, the quashing of criticism of TGiE policies nevertheless represents moments of rupture in the TGiE's performance of liberal democracy. They are moments when the rehearsal falters and international audiences who are paying close attention realise that Dharamsala does politics differently. What is understood within the community as loyalty to their revered spiritual leader can appear to Western audiences as an anachronistic mode of politics.

The scripting of the Middle Way therefore provides an important insight into internal, 'back stage' dynamics and tensions within this community (cf. Goffman 1959). However, returning to Sino-Tibetan relations, ways out of the current impasse for the Tibetan side appear limited, and some of those who support *rangzen* argue for Tibetans to say 'we tried' with dialogue, and open up other options for discussion, including reasserting claims to sovereign independence (*Tibetan Political Review* 2012). As Rabgey and Wangchuk put it:

> By turning its back on Beijing altogether, the exiled Tibetan leadership could preserve its key asset—the legitimacy conferred by the Dalai Lama's return—for a better day and a better deal. (2004: 41)

Such a strategy may offer a rejuvenated role for the TGiE in carving out a future for Tibet, giving renewed purpose to an institution that has, to date, been comprehensively excluded from all engagements with the occupying state. But where does any of this China/Tibet-facing role leave the TGiE's rehearsal of stateness?

Thus far the TGiE's performance of statecraft to the Chinese government audience has been somewhat self-contradictory. On the one hand, the actual stateness of the exile polity is a barrier to relations with China. The more self-governing the Tibetan exiles seem to be and the more 'state-like' the Tibetan government in exile appears, the more fearful China is. The view from China is precisely that the exile government *is* state-like and effective in mobilising Tibetans on both sides of the Himalayas (an impression that often exaggerates the TGiE's actual authority). Aware of this situation, the moves to downplay the TGiE's state-like functions in India, including the modified language to describe its activities, are an attempt by the exiled leadership to convince the Chinese that the Dalai Lama and TGiE are serious about the Middle Way Approach of seeking autonomy rather than independence for Tibet. On the other hand, and as noted above, the TGiE's discourses and practices of governance deliberately position it as the virtuous 'other' to communist China, certainly when presenting its case to Western audiences.

Here, China becomes a foil for TGiE's performances of stateness: where Chinese one-party rule in Tibet is presented as ruthlessly materialistic, destructive of natural environments and traditional culture and intrinsically undemocratic, so Tibetan governance in exile is premised on Tibetan Buddhist values of compassion, promoting small-scale development, organic farming and participative democracy. It could, perhaps, be argued that China is perceived and portrayed as a kind of pantomime villain by Dharamsala. It is as important as an imagined audience onto which tropes of malign governance are projected, as an 'actual' one to be engaged with in realpolitik. To an extent this is due to China being an unknown quantity for most of those born and brought up in exile. But with the TGiE's enactment of democratic governance part of a broader strategy to delegitimise the authority of the Chinese occupying state the rehearsal of stateness is also enacted and presented as a form of critique. In this guise, rehearsing the state can be read as an innovative method of resistance.

Conclusion: Ambivalent Stateness?

The purpose of this chapter has been to turn attention outwards from the domestic functioning of the TGiE to how it seeks to engage with and represent itself to key external audiences. In exploring *how* the TGiE manages its relations with the host state India, Western powers and the Chinese occupying state, attention has focused on both the performative and the discursive. What is striking is how the TGiE has developed distinct scripts for engaging with each of these audiences. The exiled leadership have taught themselves the language of international democratic governance and universal rights, and have become increasingly adept at packaging the Tibetan cause in such terms for Western audiences. More muted scripts of self-sufficiency and compliance with the laws of the land are promoted to Indian audiences, albeit calibrated according to the scale at which engagement is enacted. In recent engagement with China, the register shifts to one of religious legitimacy and conciliatory compromise on the future status of Tibet. The image of the TGiE that is projected with each of these scripts also morphs over time and depending on which audience is being engaged with. This government-in-exile projects a distinctly state-like presence in front of Western parliamentarians, presenting itself as an aspiring equal amongst democratic legislators. Yet, simultaneously, a deliberate show of *non*-stateness is enacted vis-à-vis the Indian government, at least on official levels where it plays the role of a grateful guest, not wanting to offend its host. Meanwhile in relations with China the exile government is a largely silent actor, excluded from formal dialogue with Beijing and careful to underplay its state-like qualities in the hope of dialogue resuming.

What emerges from a focus on how the TGiE is represented performatively and discursively to different audiences is a distinct ambivalence in the relationship between this polity and the concept of the state. Such ambivalence vis-à-vis the state is also apparent in the current exile leadership's visions for the political future of the homeland. For, at the same time as the Tibetan leadership is promoting its democratic and technocratic practices of good governance in the present it is also – through the Middle Way Approach – distancing itself from state discourses when narrating the *future* of Tibet and the ultimate goal of the Tibetan struggle. Many within the Tibetan diaspora were taken aback when the Middle Way was announced, and persistent questioning of this strategy – as well as outright opposition to it – has continued from those within the diaspora who disagree with the concession of a political status less than independence (*rangzen*):

> As a kid growing up in school, the Elders gave us the most wonderful dream – a dream called 'Free Tibet'... Today we are grown up and ready to fight for that dream, but the rules have changed. There is no longer that freedom to fight for. The goal post has moved... There is no glory in battling for a compromise, nor does the compromise look hopeful. (Tsundue 2004: 19)

This relinquishing of dreams of political independence are all the more vexing in light of the fact that visions of just such an independent, democratic Tibetan state were scripted in quite precise terms by the TGiE in the early 1990s in the form of 'Guidelines on Future Tibet's Constitution'. This blueprint for future governance of Tibet set out land tenure, welfare and economic systems and the rights of Tibetan citizens, and outlined a situation where the TGiE would be dissolved as soon as the issue of Tibet is resolved, and Tibetans currently residing in Tibet would head a popularly elected government. At first sight this 'roadmap' for future Tibet is easy to dismiss. It is a decidedly idealistic document that was very out of touch with realities in Tibet where, as Emily Yeh (2013) documents in her detailed study of the transformation of Tibetan landscapes through development projects, Chinese state power has been systematically consolidated over recent decades. However, despite this marked disjuncture with the situation on the other side of the Himalayas, the laying out of such an alternative political future was an important and revealing imaginative exercise in and of itself. It was an attempt to take back the initiative in terms of how Tibet's past is remembered, and, more importantly, how its future is envisioned. As Gabriel Lafitte (1999) notes, the Planning Council's drafting of these guidelines, as requested by the Dalai Lama and under the direction of the *Kashag*, was a key opportunity to draw on the experience of writing five-year IDPs in order to imagine alternative – and in many ways utopian – futures, and to debate their feasibility.[16] With the dominance of the Middle Way script for the future of Tibet the space for prefigurative debate facilitated by this exercise in formal future planning has arguably closed down. There has, of course, been a predictable fading of the dream of return as the period of exile has lengthened from years to decades. However, the fact that the nature of that future has shifted with the Middle Way and its recent modifications has created something of an existential crisis over the exile project:

> The dream of returning to Tibet, to an independent Tibet, of course it is a dream and we're not stupid, we know it's more or less impossible in our lifetimes, but we still need to hold onto this, because without it who are we and what are we doing in exile? Why are we here, why do we have this government, these settlements, why all this effort if we are not going to go back? If we are planning for exile, then we don't need to spend all this time and money and effort on our institutions, we just take citizenship and go on with our lives. It's this dream of the future which defines us and what we do. (Tibetan student, Delhi, September 2011)

If the future being currently envisaged through the Middle Way Approach is therefore fundamentally different to that being dreamt of, what then is the role of *exile* politics and what happens to the project of rehearsal? It is to such questions, and the broader applicability of the notion of rehearsal, that the following, and final, chapter turns.

Endnotes

1   The Tibetan community in South Asia has been quick to offer financial and material assistance during natural disasters in the region, including the 2004 Indian Ocean tsunami and the 2015 Nepal earthquake.

2   Tibetan performances of gratitude were on full display during a series of 'Thank You India' events in 2009 to mark 50 years in exile (e.g. http://tibet.net/2009/03/31/tibetans-mark-50-years-in-exile-express-gratitude-to-india/).

3   Tibetans holding a valid Registration Certificate are entitled to apply for an Indian government-issued Identity Certificate. This document is not a passport but it can be used for travel purposes, with visas issued on it and a 'no objection to return to India' stamp required to re-enter India (see McConnell 2013a).

4   For discussion of Tibet's political relations with foreign states prior to 1959 see McKay (1997) and McGranahan (2010).

5   Central to this has been a long history of describing the world of diplomacy through theatrical analogies, to the extent that the comparison between actor and ambassador has become something of a cliché (Hampton 2009: 143).

6   Although the Dalai Lama no longer appoints the heads of the Offices of Tibet – in line with his retirement from politics in 2011 – these individuals have thus far retained the title 'Representative of the Dalai Lama'.

7   For example, a conference on 'China's New Leadership: Challenges for Human Rights, Democracy and Freedom in East Turkestan, Tibet and Southern Mongolia' was organised by the World Uyghur Congress, UNPO and National Endowment for Democracy in Geneva in March 2013. See http://www.uyghurcongress.org/en/?p=20070

8   For example, President Obama's meetings with the Dalai Lama in 2011 and 2014 were held in the Map Room rather than the Oval Office, the usual venue for visiting heads of state.

9   The Panchen Lama – second most important incarnation in the Gelug school of Tibetan Buddhism – and the Dalai Lama have traditionally played a role in recognising each other's reincarnation. In May 1995 the 14th Dalai Lama announced, from Dharamsala, the recognition of six-year-old Gedun Chokyi Nyima as the 11th Panchen Lama. Interpreting this announcement as a direct challenge to their authority, the Chinese government denounced the Dalai Lama's decision and Gedun Chokyi Nyima was detained by Chinese security forces and has not been seen since. In November 1995, the Chinese authorities appointed their own 11th Panchen Lama, Gyaltsen Norbu.

10 This includes a positive reception amongst prominent Chinese intellectuals, and enthusiastic support from existing autonomous polities, who see it as in line with their own experiences of self-governance. See: http://www.phayul.com/news/article.aspx?c=2&t=1&id=26022&article=Resolutions+of+Autonomous+Regions+in+Support+of+Tibet

11 See: http://tibet.net/important-issues/sino-tibetan-dialogue/statement-on-sino-tibetan-dialogue-by-envoys-of-his-holiness-the-dalai-lama/statement-of-special-envoy-lodi-gyari-head-of-the-delegation-which-visited-china-and-tibet-dharamshala-28-september-2002/

12 In 1997, the Dalai Lama called a referendum of Tibetans in exile to decide on the best course of action to resolve the Tibet question. However, in a preliminary poll 64% of those balloted opted not for one of the four options outlined but expressed the opinion that they would rather follow whatever the Dalai Lama thought was best (DIIR 2010). The referendum was thus cancelled and the Middle Way Approach was effectively adopted by the TGiE by default.

13 The 2008 Memorandum was followed in 2010 by a 'Note' which addressed 'the principal concerns and objections raised by the Chinese Central Government regarding the substance of the Memorandum': http://tibet.net/important-issues/sino-tibetan-dialogue/note-on-the-memorandum-on-genuine-autonomy-for-the-tibetan-people/

14 See: http://www.cfr.org/tibet/conversation-sikyong-lobsang-sangay/p30679

15 See: http://tibet.net/2013/05/16/clarification-on-sikyongs-talk-at-council-on-foreign -relations/

16 On his retirement from political responsibilities the Dalai Lama declared that the 1992 Guidelines and 1963 Draft Constitution had 'become ineffective' (Speech to 14th Assembly of Tibetan People's Deputies, 14 March 2011: http://www.dalailama.com/news/post/657-message-of-his-holiness-the-dalai-lama-to-the-fourteenth-assembly-of-the-tibetan-peoples-deputies).

Chapter Eight
Conclusion: Rehearsing Stateness

Do not
    conjure up before us
        gruesome images of reality
            which cripple our hope
                shatter our dreams.

But leave us
    our dreams of the future
        yet to come and our
            visions of hope and realization.

Do not tell us
    'you are rootless,
        far removed from the sacred
            glory of your heritage … '

Leave us alone!
    Give us the chance to pursue
    our search for our
        shores and shapes…
    Extract from *Fresh Winds*, by K. Dhondup[1]

Rehearsing the State: The Political Practices of the Tibetan Government-in-Exile, First Edition. Fiona McConnell.
© 2016 John Wiley & Sons, Ltd. Published 2016 by John Wiley & Sons, Ltd.

Uncertain Futures in Exile: Rehearsal Unravelling?

Anxieties about the future have been a recurrent theme in my interviews and inter-
actions with exiled Tibetans in India: how and when can I get a visa to Canada/*Ari*
[America]/Swiss [Switzerland]? Are my children losing their Tibetan identity?
What shall we do with our late relative's house in Bylakuppe? What will happen if
India turns against us? Will I see my parents back in Tibet before they die? Will the
Dalai Lama return to Tibet in his lifetime? Perhaps most pervasive amongst these
concerns are those over what will happen when the current Dalai Lama passes
away. Political volatility has historically defined interregnums between the reigns
of successive Dalai Lamas, but the current situation is unprecedented. Will the
Tibetan community fragment into competing factions without its unifying leader?
Will the Tibetan freedom movement dissipate, or look to alternative, perhaps violent, strategies? Will the Indian government and public change their stance to
the Tibetans on Indian territory once such an internationally renowned leader is
no longer around? Or will the elected *Sikyong* be in a strong enough position to
provide leadership and ensure the continuity of the TGiE's legitimacy? What *is*
certain is that, with the lead playwright no longer around, the scripts for the exile
present and the hoped-for future will be increasingly up for grabs.

Linked to anxieties around a post-Dalai Lama future are concerns around the
vulnerabilities of the community's presence within India. As outlined in the previous chapter these centre on the unpredictability of India's position on Tibet
given its contingency on Sino-Indian relations, and the fact that the presence
of the Tibetan community in India is largely underpinned by goodwill and oral
understandings rather than formal legal agreements. But even if there is no substantive change in India's position, the future of Tibetans in South Asia remains
uncertain for other reasons. As noted in Chapter 4, with a lack of job opportunities the younger generation is looking overseas to North America, Europe
and Australia, with the consequence that the settlements are in seemingly termi-
nal decline. Describing this demographic transition as a 'diasporisation' of the
refugee population the Chief Planning Officer expressed his concern thus:

> In South Asia we are losing our strong base and stability. This is a major issue and a
> priority for us. The next 10–15 years will be key .... We could lose half the population
> from South Asia with migration to the West and decreasing numbers of Tibetans
> coming from Tibet. So the question is, where is next for our community? (October
> 2012)

In response to this predicament different scenarios have been mooted, albeit not
as yet actively planned for. These include an interim measure of scaling back and
consolidating the dispersed settlements, complemented by attempts noted in previous chapters to establish a 'sister settlement' programme linking with Tibetan
communities in the West and to foster transnational solidarity and social capital

exchange through the Tibet Corps programme. In the longer term a situation whereby the TGiE as an institution remains in India but the dependent population shifts to the West may be a possibility, or indeed the physical presence of the TGiE is downscaled entirely to be replaced by a networked-style structure that operates through globally scattered hubs and online forums. These would be significantly different configurations of this exile polity that, in entailing a shift from the emulation of state-like governance practices to the enactment of transnational exchanges more redolent of other diasporic communities, open up the question of how the legitimacy of the TGiE might be retained and reconstituted in such circumstances.

Such speculation is intriguing, but the fact that such future scenarios are being contemplated at all points to a broader shift in the timeframes that are currently being prioritised. Perhaps inevitably given the decades now spent in exile, there has been a notable transition from tangible expectations of return to the homeland amongst the first generation, to an acknowledgement that for those born in the diaspora, their own futures are likely to be lived in exile. As a freelance Tibetan journalist in Dharamsala put it:

> at school and when we're young we are taught to dream of returning to Tibet. In art class you have to draw what a future Tibet will be. Now I know that this won't happen in my lifetime, so now my dream of the future is a strong Tibetan run community – the horizons have changed, they've come in, they've closed in (September 2011).

In many ways, the closing in of future horizons has been precipitated by the pressures of everyday life. With uncertainties around employment and legal status a common challenge for many in the South Asian-based diaspora, the energy and resources of individuals and the TGiE itself have increasingly focused on the here and now of coping in the present rather than aspirations for longer term futures back in the homeland. Yet, seen in conjunction with the exile community being increasingly disconnected from the situation and population on the ground in Tibet, and the distinctly less-than-statehood future being promoted through the Middle Way Approach, such shifting priorities and temporal horizons have potentially profound implications. What is the present and future role of the TGiE? What is this exiled government rehearsing *for*?

The raising of such questions is not a nihilistic attempt to undermine the arguments made in this book in these final few pages. Quite the opposite. I want to suggest that the notion of rehearsal allows us precisely to focus on and to question such fundamental dilemmas of prolonged exile. It is also a provocation that encourages us to consider the myth of the state and its role in the framing of contemporary international politics in a new light. In the remainder of this concluding chapter I use the lens of rehearsal both to summarise key arguments made in the preceding pages and to suggest broader insights that the notion of rehearsing stateness can offer into the nature of the state and future configurations of geopolitics.

The Contingency of Stateness

> Opportunities for analysis expand in messier contexts, where understandings of appropriate form (what should a state look like?) and relations (how should states behave toward each other?) are contested. (Strang 1996: 23)

Central to the arguments made in this book is the assertion that the paradox of exile Tibetan state-building in a situation of statelessness can offer a useful insight into the 'norms' of conventional statehood. Anomalous polities such as the TGiE should not be seen as exotic and 'quirky' but rather as examples of spatially/temporally/legally constrained stateness, which can provide a window onto processes of state-building and practices of statecraft. As such, the ambition here has not been the production of a totalising theory of the state. Instead, following Begoña Aretxaga, the aim is 'to leave the state as both an open notion and an entity, the presence and content of which is not taken for granted but is the very object of inquiry' (2003: 395). By documenting the rehearsal spaces, roles, scripts and audiences through which exile Tibetan statecraft and stateness are daily (re)produced this analysis has, I hope, the potential to offer insights into what constitutes state practices and thus what polities have to do in order to become state-like and gain legitimacy in the international arena.

Under conventional definitions of the state (e.g. Mann 1984; Weber 1919/2009), the TGiE's seemingly insurmountable barriers to achieving recognised sovereign statehood and struggles to protect or defend its population given its restricted law-making abilities mean that this entity will never qualify as a state. But, as has been documented in the preceding chapters, despite significant legal and territorial restrictions, the TGiE articulates a range of expressions of state authority on different registers – from territorial governance (Chapter 4), to rational(-legal) authority (Chapter 5), governmentality and the establishment of a social contract with Tibetan 'citizens' (Chapter 6) and international diplomatic engagement (Chapter 7). Moreover, in attending to how the TGiE is constituted through discourses and texts (e.g. the 1991 Charter, Integrated Development Plans), everyday practices of resource (re)distribution (the payment of *chatrel*, delivery of welfare stipends) and materialities of stateness (the Green Book), this study concurs with and contributes to critical approaches that perceive the state as a 'set of practices enacted through relationships between people, places, and institutions' (Desbiens et al. 2004: 242). Following the ethnographic focus on agency and the bureaucratic messiness of state practices that critical definitions of the state espouse, the foregoing chapters argue that it is at the local level and through everyday interactions, objects and people that the TGiE is 'magicked into existence' as a state-like authority (Sidaway 2002: xi). Yet this is a case where this very everydayness of the notion of the state is brought into sharp relief by the juxtaposition of the *extra*-ordinariness of the territorial and legal situation of exile. In exceeding conventional (read 'Western') understandings of state effects, territory and legitimate authority, the TGiE thereby points to stateness operating through the medium of excess (cf. Aretxaga 2000; Navaro-Yashin 2002).

The favouring of a focus on everyday practices over linguistic and discourse analysis has been a notable trend within critical scholarship on the state and institutional ethnographies (e.g. Neumann 2012). However, though the renewed attention paid to embodied action is analytically productive, the line that this research has taken is one that troubles the distinction between the discursive and the practised. Rather, I have demonstrated how bringing together these approaches in the context of a focused case study is key to exploring the relationship between the proclaimed rationalities of government on the one hand, and the lived experiences of state(like) power on the other. As such, in offering grounded examples of *how* the state as an 'effect' is constituted, this book has brought to the fore the importance of a relational and a multi-scalar approach to understanding the state: one where attention is paid both to the material and the symbolic, practices and discourse, bottom-up as well as top-down perspectives. Illustrative of this is my reading of statecraft as the cultivation of the art of government. As such my focus has not exclusively been on elite political actors who seek to narrate nationhood but has also encompassed the everyday practices of TGiE bureaucrats at a range of ranks and the engagement of exiled Tibetan 'citizens' with *their* government. The rehearsal of statecraft thus calls to attention the embodiment, comportment and affect of state practices and questions the extent to which such performances are an individual or collective affair.

However, viewing the example of the TGiE through the lens of rehearsal does more than simply refute the state as a monolithic institution. It offers particularly valuable leverage in demonstrating the partial and processual nature of stateness and in exposing the contingent practices that underlie the social construction of political power in so-called 'normal' states (cf. Hansen & Stepputat 2001). In bringing to the fore the performative, iterative and temporally determined nature of the state as process, the notion of rehearsal – used in this book as an analytical tool and a provocation rather than an explanatory framework – thus provides an alternative and revealing aspect to the idea of stateness beyond this case. For example, *within* states the practices and narratives of political oppositions – whether formal parties/shadow cabinets or rebel/insurgent movements – have distinct resonances with the notion of rehearsal. They are 'trying out' and experimenting with different state policies, practising the roles and duties of being in government, and, in some cases, creating and promoting their own symbols and institutions of state power, but without either the democratic mandate or legal recognition and responsibilities. There are certainly parallels here with the TGiE, but the notion of rehearsal also has relevance in the realm of legitimate de jure governance. If we are to accept that 'conventional' states are always in process, then it perhaps does not require too much of a conceptual leap to argue that all states are to some extent rehearsal states, experimenting with political discourses, practices and materialities in order to evolve and adapt to changing global geopolitical realities from the global financial crisis to the threat of ISIS/Daesh and future uncertainties over the impacts of climate change.

Rehearsal thus not only offers a useful insight into the state as in a continual situation of emergence (Jones 2012), but attending to rehearsal as performative – holding in tension dramaturgical performance (Goffman 1959; Turner 1982) and performativity as reiterative (Butler 1990) – also exposes the *unevenness* of stateness. Performances of the TGiE's state-like governance are more visible in some spaces (exile institutions and settlements) than others (scattered communities and the diaspora in the West) and are also articulated by various constituencies in different ways and to distinct audiences. As noted in the previous chapter the TGiE's stateness is performed in markedly different ways – and to different degrees – to its host state, occupying state and Western audiences to which it turns for political and material support. Meanwhile Tibetans in Tibet are barely part of the rehearsal at all – there by proxy in the electoral system and in the TGiE's claims to legitimacy as representatives of all Tibetans – and within the diaspora some Tibetans are sidelined (e.g. 'newcomers' from TGiE jobs) whilst others (e.g. disaffected youth) self-exclude from the exile project.

Indeed, if practice, training and developing expertise form core components of the participative aspect of rehearsal then the counterpoint to participation – apathy – is also always potentially present. As documented in the preceding chapters, in a situation of prolonged exile sustaining participation in the broader project is an increasing challenge. As such, belief in a 'script' – a series of narratives that underpin the exile political project – in the playwright and in the possibility of there *being* a final performance, becomes key. The inherent vulnerability of this belief and the need to continuously foster it provides a revealing angle on the construction of political legitimacy and the nature of the social contract being articulated both in this case and in polities more generally. This does not, however, necessarily entail universal and absolute belief. As Yael Navaro-Yashin argues, drawing on Žižek (1995), contemporary state power needs cynical subjects to maintain itself: cynicism 'is an approach that reproduces the political by default' and thus 'reinstates the state over and again' (Navaro-Yashin 2002: 5, 159). In the case under examination here, where a distinct suspension of disbelief is required in the first instance to conceive of a state-like polity being rehearsed in exile, the argument that individuals carry on with 'everyday practices *as if* the state were a unity' (*ibid* 2002: 161) takes on a heightened quality. Yet because the rehearsal of stateness in exile is a deliberate and self-conscious political project there is a fine line to be trod with regards to belief and cynicism: there is a constant danger of the TGiE being seen as 'not the real thing' by its 'citizens'. Rather than cynicism acting to lock citizens into 'the chains of statism' (*ibid* 2002: 159), here cynicism can potentially threaten the very stateness of the TGiE. Therefore, unlike micro-nations and virtual states such as the Conch Republic and the Principality of Sealand whose very existence is, in many cases, an exercise in humour, social reform or self-aggrandisement (Grimmelmann 2012; Steinberg & Chapman 2009), the TGiE is dependent on its being taken seriously.

As such, more than simply the accrual of competence in governance, the TGiE also seeks to *demonstrate* competence: this is an administration that seeks to govern in the present as well as prove its credentials for the future. We might think, therefore, of parallels between this case and the practices of constituent assemblies, such as those in pre-independence India (1946–47), Italy after the Fascist defeat during World War II (1946–1948) and Nepal following the abolition of the Nepalese monarchy in 2008. These are moments when polities are formally constituting themselves as states and, as such, the case of the TGiE brings to the fore not only the practices of political consciousness-raising involved in such projects but also the fictions built up around temporalities, with those involved knowing that the process will take far longer than proposed, but still demonstrating faith in the original timeframe and the seriousness of the endeavour itself.

The Creativity of Liminality

Pausing to consider the relationship between the exile present and imagined futures in more detail, the future-oriented practices of both anticipatory logics (Anderson 2010; Collier 2008) and prefigurative politics (Cornell 2011; Vasudevan 2015) have been posed in this book as offering points of intersection and divergence vis-à-vis the case of the TGiE. Firstly, the Tibetan case brings questions of nationalism to the fore, with Tibetan national identity shaping political actions in the present, and informing hopes for the future. There are resonances here with postcolonial critiques of the notion of homogeneous, empty time of nationalism and the nation (Chatterjee 2005; Gupta 2007; Krishna 1992; cf. Anderson 1991). What I have traced in this book are a range of mechanisms through which the exile leadership has been *aspiring* to universal ideas of nationalism and citizenship, and a standardised, Western liberal conception of democratic politics. Yet, these 'ideals' are continually being undercut by the TGiE's everyday practices of governmentality, by the persistence of regional identities, and by the continuity of religious practices and principles in exile politics that are premised upon different modes of temporal ontology.

Secondly, unlike conventional prefigurative politics the TGiE is neither seeking to reject centralised authority nor challenge hierarchy.[2] In fact, quite the opposite: it is an attempt to enact a future society that in many ways resembles the status quo of the modern nation-state (discussed below). Yet, at the same time, attending to how future planning is conceptualised and plays out in an exiled Tibetan context extends our gaze beyond Western liberal democracies that have been the focus of extant work on anticipatory action. It foregrounds how Tibetan modalities of time based around cyclicity and impermanence intersect with how futures are awaited and acted upon, and what happens to anticipatory practices of imagination and calculation when the timeframe is extended seemingly indefinitely.

This is not, however, to imply a teleology to this polity or my interpretation of its functioning and raison d'être. Rather it is the multiple, overlapping and at times contradictory temporalities that were described by exiled Tibetans and TGiE officials that underpin much of how exile politics and governance is articulated. As Partha Chatterjee notes of the postcolonial world, there is 'the presence of a dense and heterogenous time' (2005: 928). On the one hand timeframes have, thanks to the five decades in exile and the lack of an imminent resolution to the political status of the homeland, become both stretched and suspended. Echoing Navaro-Yashin's reflections on the state of 'permanent disruption' that pervades the de facto Turkish Republic of Northern Cyprus where '[T]he frozen time, the time interval, has become the consciousness of temporality in this bisected space' (2003: 121), these are temporalities quintessential of the limbo of waiting (see also Jeffrey 2010). Yet, on the other hand, the urgency of the broader Tibetan cause and dreamed-of futures slipping away is also always present. 'Time is running out for Tibet' is a slogan often repeated by Tibetan activists, both old and young, and is a call to action increasingly linked to the Dalai Lama's progressing years and the fear that he will not be able to set foot in Tibet again.

As Yossi Shain notes, the paradoxical desires of 'wishing to force the future into the present while at the same time trying to preserve the past' (1989: 31) are common across exiled communities. The effective 'squeezing' of the imperatives of present-day life in exile is certainly evident in the Tibetan case, from the practical shortcomings of the symbolic electoral system (Chapter 4), to decisions to retain the status of 'stateless foreigners' in India (Chapter 6). However, it is not just in 'exceptional' situations of exile that such temporal dilemmas are apparent. The competing temporal imperatives outlined in this study also speak to the notion of the constant state of emergency in (neo)liberal democracies whereby the present effectively gets diminished in the drive to plan for the future (Adey & Anderson 2012). Again the notion of rehearsal offers insights here. As an open-ended practice rehearsal does not presuppose particular futures: the 'final performance' is anticipated but not inevitable. Moreover, each time something is rehearsed not only are the actions and intonations subtly different, but also the notion of what the final performance will be can incrementally shift. And, like the party game of Chinese whispers, the more you rehearse the greater the possibility that the final performance is different than originally envisioned. Extend this rehearsal over decades, and in the context of dynamic geopolitical relations, and it is little wonder that visions for the future of the Tibetan homeland have morphed over time (see Chapter 7). Cross-cutting these characteristics of rehearsal is the question of its very durability. The shifting temporal horizons of the exile community detailed at the start of this chapter – towards a focus on planning for *exile* futures – raise the possibility of this rehearsal of stateness having run its course. The question thus arises as to whether rehearsal itself is always somehow time limited?

The tension between prolonged waiting and anticipating futures may well be a contributing factor to rehearsal 'running out of steam', but it also throws up conceptual challenges of understanding coexisting temporalities. Indeed, to follow Partha Chatterjee's postcolonial analysis of the temporality of modern nationhood, to what extent are we seeing 'the presence of a dense and heterogeneous time' (2005: 928) as the product of displacement itself? It is perhaps a stretch too far to claim that the notion of rehearsal facilitates the making of ontological claims on the nature of temporality, but what rehearsal does do is to bring to the fore the intersecting and entanglement of temporalities as an effect of a set of practices: in this case *state* practices. In turn, the diversity of temporalities opens up productive lines of enquiry regarding the state more generally: what is the duration of different state effects? What are fleeting and what are durable? Different rhythms (cf. Lefebvre 2004) to the processes through which state effects are generated have been traced in this study, from the calculated and prioritised planning of the IDPs to the more improvised everyday interactions between TGiE officials, Tibetan 'citizens' and Indian bureaucrats (cf. Jeffrey 2013). As such, paying attention to the multiplicity of temporalities – and spatialities – has the potential to deepen the idea of state effects and to trouble the distinction between the state existing and the state emergent.

Indeed, just as temporality has been a key motif running through this book, so has space. It goes without saying that a key constraint on the ability of an exiled community to remain unified, sustain a freedom movement and make claims to legitimacy is precisely their geographical dispersal. As Edward Said remarked of the Palestinian diaspora, 'the fact that most of our people were dispersed exiles, [is] a condition in which geography became our main enemy' (2001: 546). What is striking in the Tibetan case (and one that has distinct parallels with the Sahrawi Arab Democratic Republic, or SADR), are the efforts made to appropriate 'geography' in order to facilitate a mode of territorialised governance. As such, the TGiE as it has existed since 1960 is a territory-*less* not a *non*-territorial polity. For, whilst denied jurisdiction over territory in the homeland and in exile, the TGiE's state-like functioning in India is nevertheless in many ways contingent on its de facto control over the series of Tibetan settlements and the efforts to territorialise its practices and authority (Chapter 4). In turn, the very out-of-placeness (Cresswell 1996) of this exiled polity challenges and disrupts the assumption that territory and the state map onto each other: an assumption that underpins conventional understandings of politics at a range of scales. More precisely, the exile Tibetan case encourages a rethinking of the necessary territorial configurations of state practices, performances and narratives.

As noted in Chapter 4 the TGiE's establishment and institutionalisation of nested hierarchies of bureaucratic authority, standardised working practices and integrated welfare services speaks in revealing ways to Joe Painter's (2010) work on territoriality as an aspect of the state's structural effect. Not only does it reaffirm the assertion that the process of territorialisation produces state territory, but it also provides a grounded example of 'strategic territorialisation' (Boudreau 2001) whereby the TGiE appropriates notions of territory for tactical political goals. Reflecting the observation that liminality – in this case when conventional territorial and legal structures are suspended – can produce intense creativity (Thomassen 2012; Turner 1987b) the roles of symbolism and innovation have taken on heightened importance in this case.

For example, as discussed in the previous chapters, symbolism – in the form of the parliamentary electoral system, the role of the judiciary and the values attributed to the Green Book – plays a key function both in facilitating the enactment of governance in constrained legal and territorial circumstances and in providing connections to the homeland of Tibet. This employment of symbolism has interesting parallels with the role of symbolic and ritual authority in enabling the pre-1959 Ganden Phodrang to function over a vast territory and without a comprehensively regulatory system (Samuel 1982, 1993). There are also resonances with other exiled polities such as the SADR based in Algeria and Palestinian communities in Lebanon, where placenames in the homeland and dates of key national events feature prominently in 'state' institutions and spaces (Beker & van Oordt 1991; Hanafi 2008; Mundy 2007; San Martin 2010). Of course symbolism also plays an important role in the discursive construction of more conventional states. Anthropologists have documented '*symbolic representations* of the state as a locus and arbiter of justice and a symbol of larger society' (Hansen & Stepputat 2001: 10, emphasis in the original) with, for example, Akhil Gupta (2012) outlining how the Indian state employs a range of cultural practices to symbolically present itself to citizens. But what is striking in the case of the TGiE is the instrumental and strategic role that symbolism takes on vis-à-vis community unity and circumventing the limitations of functioning in exile, and the tensions that arise when symbolism is overplayed. Seen most clearly in the tension between symbolic territory and local political concerns in the context of the Tibetan Parliament in Exile electoral system, this case thus raises pertinent questions about the practicability of stretching of political authority and participation beyond 'national' frontiers, and how this affects issues of legitimacy and the meaning of political representation.

The puzzle of legitimacy in the absence of legality has been a recurrent theme in this study. For, whilst international recognition of the exile Tibetan government has never been forthcoming, is not currently being sought, and is unlikely to be granted anytime soon, nevertheless, remaining outside the formal inter-*state* system does not preclude the exile administration's claims to legitimacy nor Tibetans' endorsement of it as their legitimate representative. As the foregoing chapters have detailed, the TGiE's claims to and ongoing construction of legitimacy are key defining features of this polity and, as such, the reading of legitimacy presented here is one of a dynamic and contested process rather than an end goal that a polity must 'secure'.

There is therefore an ambiguity at the heart of the TGiE's constructions of legitimacy in exile. This is manifest in two way. First is the TGiE's own ambiguous status: it is sometimes perceived as a glorified *non*-governmental organisation, a provider of welfare for Tibetan 'refugees' and the lead campaigner for the Tibetan cause. However, following Feldman's assertion that 'civil service authority partly entails carving out a space that distinguished its practice from other sorts of practice' (2008: 64), it is the exile Tibetan polity's purposeful performance of stateness through state-like bureaucratic structures, practices and the training of its staff in the arts of statecraft that in effect positions this polity in the realm of stateness as distinct from NGOs. Second, legitimacy is in itself an ambiguous concept. Yet, rather than see such ambiguity as grounds for critiquing the utility of this notion (e.g. Shain 1989), as Alice Wilson and I have argued, the polysemous relationship that legitimacy has to legality and to the state not only makes it accessible to polities like the TGiE, which lack full legality, but also the political, legal and territorial liminality of this case provides opportunities for creative experimentation in producing governmental legitimacy (Wilson & McConnell forthcoming). For example, given that the TGiE's authority cannot be based on legal powers and it thus cannot *enforce* national loyalty or compliance through purely coercive means, the exiled leadership has instead focused its efforts on asserting a significant degree of societal pressure and moral authority through its control over the education system, the granting of citizenship and providing for its 'citizens' materially as well as the powerful charismatic authority of the Dalai Lama.

In particular, it is the quotidian bureaucratic functioning of the TGiE within seemingly exceptional situations and spaces that reinforces this polity's claim to political legitimacy, and it is the state-like-ness of this bureaucracy that goes a long way in providing an 'illusion of stability and continuity' (Ferguson & Mansbach 1996: 401). As traced in Chapter 5, the TGiE's establishment and maintenance of a functioning and distinctly state-like bureaucracy has notable parallels with Ilana Feldman's analysis of the role of quotidian bureaucratic practices in providing stability of rule in what were politically volatile periods of British and then Egyptian rule in Gaza in the early and mid-twentieth century.[3] In terms of the contemporary period, the case of the TGiE has the potential to provide a pragmatic model of durable self-governance in exile – a model whereby refugees preserve their culture and identity and the burden on the host state is significantly reduced over time – which is thus distinct from conventional refugee policies of integration, assimilation or repatriation. The TGiE's provision, to date, of a degree of security for this refugee community, its de facto autonomy within mutually agreed legal boundaries, organised networks of cultural and educational institutions, stateness without statehood and practising of democracy could be a valuable template for other refugee communities, or even for indigenous populations and national minority groups. As such, this case has the potential to contribute to debates in refugee policy regarding the issue of 'durable solutions' (Van Hear 2003). This has particular resonances with uncertainties around the political futures of small island states such as the Maldives and Tuvalu, which face threats to the existence or the habitability of their territories due to rises in sea levels induced by climate change. The innovative territorial arrangements and governance strategies of the TGiE and its negotiated tacit sovereignty within the host state of India may provide pointers for future scenarios of ex situ nationhood (Burkett 2011; see also Blitz 2011; McAdam 2010; Yamamoto & Esteban 2010).

The State (Idea) as Aspirational

> The redeeming thing about exile is that when your 'old world' has vanished you are suddenly given the chance to experience another... Indeed, what you eventually get is not just a 'new world,' but something philosophically more consequential: the insight that the world does not simply exist, but it is something you can dismantle and piece together again, something you can play with, construct, reconstruct and deconstruct. (Bradatan 2014: n.p.)

Having set out *how* the TGiE has sought to emulate state practices, I now want to turn to the question of *why* this exiled government has put such work into emulating this particular form of political organisation and, more broadly, what this tells us about the influence and durability of the state idea. For, by arguing that stateness is a process that is always partial and contingent, I do not want to downplay the ideological influence of statehood itself. Rather, though the exile government cannot create a reified state through disciplinary mechanisms, its selective appropriation of aspects of statecraft is central to demonstrating its ability to govern competently and to claiming legitimacy. The signifiers and institutions of statehood can thus be seen as strategic resources that are continually deployed by both recognised and unrecognised polities.

In his persuasive discussion of the sovereign state system as a 'political-territorial ideal', Alec Murphy argues that 'the modern territorial state has co-opted our spatial imaginations. And the co-opting has been so far-reaching that we accept it unproblematically' (1996: 107). Even in an era when a plethora of non-state political actors and institutions vie for space on the international state, the statist paradigm continues to dominate. Indeed, it is precisely for those communities who have thus far been denied statehood that the power of the state as an idea and an ideal has particularly compelling resonances. As Tozun Bahcheli and colleagues note, the appeal of sovereign statehood 'has ignited the ambitions of scores of societies for whom such a status would have once seemed both absurd and unreachable' (Bahcheli et al. 2004: 1). This is readily apparent in the state-like operations, claims and pretentions of ISIS/Daesh, Kurdistan,

Scotland and Catalonia to name but a few examples dominating contemporary media headlines.

As the 'great enframer of our lives' (Hansen & Stepputat 2001: 37) the myth of the state is thus a powerful and enduring one. The fact that the TGiE is currently keeping a discursive distance from demanding statehood or even claiming to act in state-like ways in its dealings with China is evidence of Aretxaga's assertion that 'state continues to be a powerful object of encounter even when it cannot be located' (2003: 399). Indeed, the idea of the state is also a basis for social action in cases where the state is ostensibly absent. As Morten Nielsen discusses in the context of peri-urban areas of Maputo, Mozambique, 'even when the reality of state dysfunction is widely accepted, "ordinary people" continue to invest themselves in... ideas [of the state]' (2007: 695). The state may, as Abrams (1988) argued, be elusive and illusory, but as an aspiration it has powerful effects.

In paying analytical attention to the enduring appeal of the state as an ideal to aspire to the state needs to be seen as far more than simply what is contained within defined borders. The idea of the state also engenders a sense of political purpose, and the institutional materiality of state-like functions have important affective impacts on communities (Navaro-Yashin 2012; Stoler 2007). What I have argued here is that the example of exile Tibetan politics suggests an affective relationship to the state based on hope and aspiration. This is therefore a mode of affect that not only runs counter to tropes of fear, suspicion and terror that are often associated with the state (Arendt 1976; Bauman 1991) but also acts as a counter-narrative to the harrowing reports of human rights abuses, patriotic re-education of monks and nuns, and forcible settlement of nomads (Human Rights Watch 2010, 2013) that dominate reports coming out of contemporary Tibet. In bringing to the fore the hopeful expectations of exile Tibetan politics this is in no way to detract from the situation inside Tibet, but rather to note that the enactments of stateness in exile have been a strategic lifeline for what, on paper, is a quintessential 'lost cause': a cause where'[T]he time for conviction and belief has passed... [and it] no longer seems to contain any validity or promise' (Said 2001: 527).

What the everyday stateness of the TGiE has enabled is the prolongation of belief in the Tibetan cause and wider freedom movement. It has provided a focus point for the diaspora, a 'way of keeping the community together and united, giving us a purpose' (monk, Dharamsala, September 2011). Central to that purpose has been the fostering of aspirations towards a political future different to the past and the present in the homeland. As noted in the previous chapter the sustaining of this aspiration appears to be faltering in light of contested visions for what a future back in Tibet might look like and a growing disjuncture between exile politics and the realities on the ground in the homeland. As such, we might pose the question that Tracy Davis asks of the protracted civil defence exercises: 'what if the techniques and ontology of rehearsal creates a sense of false security?' (2007: 330).

This challenge notwithstanding, the politics of hope that the rehearsal of stateness in exile generates has the potential to contribute to two sets of debates. First are discussions around the political possibilities of exile. Mindful of avoiding a redemptive reading of exile (Said 1984), this study has demonstrated how the situation of displacement can, under certain conducive circumstances, foster the building of nations, innovation of political ideas and envisioning of different futures. As such, prolonged exile can itself be a type of social and cultural resource as well as a frustrating and disillusioning experience. The 'freedom of exile' is a freedom to experiment and be politically innovative in ways that are impossible under conditions within the homeland. Just as 'Manchukuo served as a laboratory for what was not possible to achieve within Japan itself' (Duara 2003: 250) so the Tibetan settlements in India provide spaces for rehearsing modes of democratic politics that are proscribed inside Chinese-governed Tibet. This may not be experimentation in radically new modes of politics as seen, for example, in contemporary Guantanamo Bay (Reid-Henry 2007) or in the inter-war experimental states (Roslington 2013) or, indeed, in articulations of subaltern politics more generally (hooks 1990). Indeed, in many ways the TGiE is emulating precisely the status quo: liberal democratic statehood. Nonetheless the TGiE's rehearsal of stateness has proved to be an effective mechanism through which political agency is claimed. The time and space for experimentation offered by exile has facilitated ways of ensuring that the meaning and renewal of the political is maintained, and enabling the capacity to act in the face of uncertainty. These are achievements that views of exile as solely alienating and debilitating overlook.

The second set of debates around affect that this case speaks to is the small but growing body of work on the relationship between the state, affect and political aspirations. Not only can it be asserted that the 'type of political control exercised by the state, as described by Weber and others... lies as much in the realm of *political aspiration* as it does in political practice' (Jones 2007: 4, emphasis added), but narratives around hope, reverence and desire that can be generated around the idea of the state are also revealing (Nuijten 2003). The case of exile Tibetan stateness corroborates the notion of the state as a 'screen for political desires and identifications' (Aretxaga 2003: 393) that a number of anthropologies of the state put forward (e.g. Navaro-Yashin 2002; Stoler 2007). The idea of the state here is as an instrument not of domination but of empowerment and emancipation. On one level an enchantment with the idea and ideal of the state can be traced. Resonating with Navaro-Yashin's assertion that the concept of fantasy 'enables us to study the enduring force of the political' (2002: 4), exiled Tibetans' belief in the TGiE as their legitimate representative and as a functioning 'state' is essential to this polity's continued claims to existence. But as documented in this book, individual Tibetans in the diaspora do not cling to this state-like polity for purely pragmatic reasons. Indeed, in many instances the fantasy of the TGiE is vulnerable to exposure precisely because this polity cannot legally defend its 'citizens', guarantee economic conditions or provide law and order. Rather exiled Tibetans believe in this 'state' for other reasons: in order to keep the cause of Tibet alive, and as a mode of resistance. Indeed, rehearsing stateness in this case is also a form of critique and resistance, with the TGiE's performance of democratic governance a key component of the broader strategy of delegitimising the authority of the Chinese occupying state. What we see here, therefore, is the state not only as fantasy, but as strategy. There is a mutual co-dependence between the idea of state effects (Mitchell 1991) and state affect.

A key element of the TGiE's strategic rehearsal of statement is the mimicking of democratic governance in order to seek approval from and prove its legitimacy to Western audiences. The exiled Tibetan leadership has sought to replicate rational, structured state authority, controlled from Dharamsala and bearing the key hallmarks of a liberal democracy. In other words a model of the state in which nationalism provides a key basis of legitimacy and where state power is centralised and bureaucratised. Read through the lens of Bhabha's work on mimicry whereby mimicry 'must continually produce its slippage, its excess, its difference' (1984: 126), this emulating of the structures, practices and ideologies of an 'ideal' of liberal statehood raises pertinent issues. The fact that the TGiE's emulation and appropriation of the governmental practices of recognised sovereign states – both its host India and Western democracies – occurs alongside its incorporation of traditional Buddhist modes of politics highlights the hybridity of this polity. The TGiE is not a cultural and political 'other' to the conventional state form, but at the same time it is also experimenting with and adapting culturally specific modes of statecraft. As such, this case affirms the postcolonial argument that we need to shift our vantage point from 'viewing the rest of the world as peripheries or sites for testing models crafted in the West' (Ong 1999: 24) to learning from 'diverse political contexts' (Robinson 2003: 648).

Rehearsing Geopolitical Futures: Ambivalence and Openness

The new post-Cold War global order... is not... crisply divided into entities which do and do not count as 'states'. It consists instead of a mass of power structures which, regardless of formal designation, enjoy greater or lesser degrees of statehood. (Clapham 1998: 157)

The desire of the exiled Tibetan leadership to emulate official statehood whilst simultaneously affirming a distinctively Tibetan articulation of governance also brings to the fore the notion of ambivalence. In relation to the Tibetan case specifically, ambivalence has arguably defined relations to the state in this region. As outlined in Chapter 3, statehood has long been elusive in the Tibetan context, and narratives around Tibetan history are characterised by polarised contestations

over territorial claims and religious and secular authority. The current administration's compromised vision for the future of the homeland in the form of the Middle Way Approach can be read as a continuation of this (geo)political uncertainty
(see Chapter 7). Moreover, with its simultaneous anticipation of better futures
and disillusionment of being stuck in limbo, the out of place-ness of exile itself is
defined by an affective ambivalence. It is precisely the tension between the familiar and the unusual, the banal and the exceptional, the mimicking of the status
quo and the enacting of alternative governance that is at the core of the TGiE
that, I want to suggest, offers an opportunity for exploring alternative modes of
formal politics.

On the one hand the notion of ambivalence in many ways offers a useful framing for describing what the TGiE *is*. For, as this book has sought to document, this
is a polity that is simultaneously state-*like* and state*less*. It has state-like qualities,
yet is territory-less; it is a polity that is intended to be temporary but is becoming increasingly settled; it combines nationalism (the assertion of belonging to a
place and community) with exile (the removal from and absence of such a community); and its 'citizens' also self-define as 'stateless refugees'. In temporal terms
the case of the TGiE challenges the presumed correlation between statehood and
permanence, and statelessness and temporariness. It is a polity that has increasingly become 'stuck' in exile and is far from the fleeting or temporary phenomena
of 'conventional' stateless communities. Teleological assumptions that this polity
is at a 'halfway house' on the road to statehood (Gottlieb 1993: 32) should therefore be treated with caution as there 'is no "end of history" here – no one road to
statehood' that the TGiE will ultimately find itself on (Kingston 2004: 7).

As a polity that thereby holds in tension aspects of both statehood and statelessness this case enables a progression 'beyond profitless debates as to who are
and who are not significant actors in world politics' (Hocking 1999: 21). In other
words, by blurring the boundaries of 'traditional' definitions of state and non-
state players, the TGiE collapses conventional notions of the official and proper
conduits of statecraft. However, rather than calling for a dismantling of the state,
such appropriation of these same forms of representation reconceptualises issues
of agency and actorness and highlights the pragmatic and heterogeneous constructions of international space. As such, I want to suggest that the ambivalence
at the heart of the TGiE's rehearsal of stateness should be embraced as an opportunity for considered reflection on the nature and the future (re)configuration of
political space and authority more generally. There are important parallels here
with Scott Pegg's discussion of the role of de facto states in the contemporary
interstate system, wherein he notes:

> By its very nature, the de facto state is well suited to situations where the interna-
> tional community needs to be seen to be upholding cherished norms, while at the
> same time it finds creative or ad hoc ways to get around those very same norms.

Its inherently nebulous status has the additional benefit of not precluding any other future settlement arrangements. (1998: 201)

Likewise, rather than perceiving the TGiE as a challenge to the state system, it is perhaps more productive to see its ambiguous status as 'part of a cyclical return to a more diverse international system' (Pegg 1998: 231). This is not to envision a world without states per se, but rather the re-emergence of older forms of political pluralism constituted of diffused authority, overlapping jurisdictions and asymmetrical constitutional arrangements (Keating 2001). As such, the rehearsal of stateness by communities conventionally excluded from the inter-state system thereby offers a valuable glimpse of possible geopolitical futures. Of, for example, a more disparate international society characterised by varying degrees of stateness, sovereignty and territoriality (Agnew 2005; Murphy 1996), which, in turn, should encourage a more general shift towards 'thinking of the state, sovereignty and territory in the plural rather than the singular' (Anderson 1996: 135).

Christopher Clapham's notion of degrees of statehood is instructive in this regard. Drawing on his research on African guerrilla movements, Clapham argues that when statehood is perceived as relative, different entities can be seen to meet 'the criteria for international statehood to a greater or lesser degree': conventional states often fail to enact the range of statehood functions, and less-than-state entities take on 'attributes customarily associated with sovereign statehood' (1998: 143).[4] Given that no state fully realises the modern state ideal (no state has a true monopoly on physical force within its territory nor restricts its power to the space within its borders), it follows that non-state entities that claim aspects of state power are players rather than pretenders who, much like 'real' states, leverage *aspects* of sovereign authority as they engage in the global political arena. By extension, if there can be degrees of statehood, then might there also be 'degrees of legitimacy' (Caspersen forthcoming)? The latter might not necessarily be a product of the former, but perhaps degrees of legitimacy might project and underpin degrees of statehood. As seen from the case of the TGiE, a disaggregated geography of legitimacy would also have a complex relation to temporality as not only is the past reworked either through alleged continuity or rupture, but also claims to legitimacy are staked on a notion of desired futures.

But what might the (re)pluralisation of our understanding of political space and the nature of stateness mean for theorisations of geopolitics? On a pragmatic level, the continued existence and rehearsal of polities like the TGiE could provide 'new ways of coping with the present, post-sovereign order' (Keating 2001: ix). As the contemporary complexity of world politics increasingly fails to correspond to the idealised model of an inter-state system this plurality of degrees of sovereignty could be 'a messy solution to a messy problem' (Pegg 1998: 194) and thus lead to greater stability in the international system. More conceptually, the notion of rehearsing stateness has the potential to be a means by which new theoretical explorations can be launched. It is an increasingly acknowledged truism that international law has not kept pace with changing geopolitical realities. Whilst acknowledging the necessity of tightly defined legal semantics in order to ensure that rulings are implemented, the inflexibility of legal terms, principles and declarations is a rigidity that has governed both what can be asked for and what can be granted. In contrast, by accommodating more varied forms of geopolitical arrangements the notion of rehearsing stateness has the potential to enrich our empirical vocabulary (cf. Gottlieb 1993). Following Camilleri and Falk, this is not a call to predict the future,

> but rather to see more clearly the direction in which it might or could be shaped. This is not a recipe for prediction or prescription but an invitation to an ethically sensitive evaluation of future possibilities. (1992: 252)

This would entail an acknowledgement of the role of contingency in the untidy discursive production of international recognition and legitimacy, an embracing of the multiplicity of stateness, an opening up of alternative geopolitical visions and possibilities, and a revaluing of the political.

Jo Sharp notes that 'this is exactly the time where a more ambitious geopolitical imagination is required' (2013: 27) and, in turn, I want to suggest that the TGiE might be representative of an era in which states need to rehearse more. An era in which numerous claims to statehood are being made, in which economic and political crises are often translated into knee-jerk reactions of so-called 'normal' states, and thus in which a more reflective approach to the nature of statecraft is required. Applying the idea of rehearsal to conventional states therefore foregrounds a number of issues: the relative weight of territory in the legitimising of state authority (compared to, for example, discourses of cultural authenticity, moral coercion or everyday governance practices); the training and experimentation that go into state practices; the importance of engendering belief in the performance of statehood; and the role of audiences (from citizens to other governments and international organisations) for the functioning of all polities. In suggesting that the state be left as an 'open notion' (Aretxaga 2003: 395), the case of contemporary exile Tibetan politics also points to the importance of futures being conceived of as open. In the words of Doreen Massey:

> Only if we conceive of the future as open can we seriously accept or engage in any genuine notion of politics. Only if the future is open is there any ground for a politics which can make a difference. (2005: 11)

With the Middle Way Approach dominating formal exile politics, prefigurative thinking may be presently sidelined within the TGiE but it has far from disappeared within the diaspora. Online forums have proved fertile ground for 'what if ... ?' discussions with, for example, the *Tibetan Political Review* initiating a public conversation on the practical issues that would have to be considered for 'Tibet's Day After'.[5] And, back in Dharamsala, alternative futures are also being scripted in more rudimentary and more performative ways:

> *On Tibetan Independence Day, 13 February, activists from Students for a Free Tibet–India set up a stall outside the TCV day school in Dharamsala. A handwritten sign reading 'Mission #freetibet From Dhasa to Lhasa' is pasted to the school wall in front of which is a small wooden table. On the table is a 'Tibet telescope', 'invented' in Dharamsala and constructed from black cardboard. Tibetan passers-by are invited to step forward, look through the lens of the telescope and 'see their future'. And what do they see? A diorama showing the Potala Palace, the main residence of the Dalai Lamas in Lhasa, with a Tibetan national flag proudly flying in front of it.*

Endnotes

1   K. Dhondup was a poet, historian and writer (for an obituary by Jamyang Norbu see: http://www.rangzen.net/1995/05/15/writer-and-historian. *Fresh Winds* features in the anthology of Tibetan poetry *Muses in Exile* (2004) edited by Bhuching D. Sonam.

2   The desire (though not fully enacted) to do away with both religion dominating politics and the central role played by aristocratic families is arguably politically 'radical' in the Tibetan context, as is the introduction of participatory democratic politics.

3   As noted in Chapter 5, unlike the Gazan case where the question of legitimacy was deliberately held in abeyance (Feldman 2008), here the TGiE is actively seeking to construct and project its legitimacy as the representative of the Tibetan nation.

4   It should be noted that dominant scholarship on sovereignty and statehood in the African context, which frames it through the discourse of lack, has been critiqued for a tendency to internalise a hierarchy that reproduces a colonial narrative about levels of modernity and development (see Sidaway 2003).

5   *Tibetan Political Review* 19 October 2011: https://sites.google.com/site/tibetanpolitical review/articles/tibetanfreedomandthedayafter

References

Abrams, P. (1988). Notes on the difficulty of studying the state (1977). *Journal of Historical Sociology* 1, 58–89.

Addy, P. (1994). History: economics and ideology. British and Indian strategic perceptions of Tibet. In R. Barnett (ed.), *Resistance and Reform in Tibet*. Bloomington, IN: Indiana University Press, pp. 15–50.

Adey, P. & Anderson, B. (2012). Anticipating emergencies : technologies of preparedness and the matter of security. *Security Dialogue* 43(2), 99–117.

Agamben, G. (1995). We refugees. *Symposium* 49(2), 114–119.

Agier, M. (2002). Between war and city: towards an urban anthropology of refugee camps. *Ethnography* 3(3), 317–341.

Agnew, J. (1994). The Territorial trap: the geographical assumptions of international relations theory. *Review of International Political Economy*, 1(1), 53–80.

Agnew, J. (1998). *Geopolitics: Re-visioning World Politics*. London: Routledge.

Agnew, J. (1999). Mapping political power beyond state boundaries: territory, identity and movement in world politics. *Millennium: Journal of International Studies* 28(3), 499–521.

Agnew, J. (2005). Sovereignty regimes: territoriality and state authority in contemporary world politics. *Annals of the Association of American Geographers* 95(2), 437–461.

Aiming, Z. (2011). The liberation of Tibet is a splendid chapter in the unification of the Chinese nation: an interview with Mr Zhu Weiqun, Vice Minister of the United Front Work Department of the Central Government of China. *China's Tibet* 2011.3 Vol. 24, 4–13.

Aldecoa, F. & Keating, M. (eds) (1999). *Paradiplomacy in Action: The Foreign Relations of Subnational Governments*. Portland, OR: Frank Cass.

Alexander, J.C. (2011). *Power and Performance*. Cambridge, UK: Polity Press.

Allen, J. (2003). *Lost Geographies of Power*. Oxford: Blackwell.

Amrith, S. (2011). *Migration and Diaspora in Modern Asia*. Cambridge: Cambridge University Press.

Anand, D. (2000). (Re)imagining Nationalism: identity and representation in the Tibetan diaspora of South Asia. *Contemporary South Asia* 9(3), 271–287.

Anand, D. (2002). A guide to little Lhasa in India: the role of symbolic geography of Dharamsala in constituting Tibetan diasporic identity. In P.C. Klieger (ed.), *Tibet, Self, and the Tibetan Diaspora: Voices of Difference*. Leiden: Brill, pp. 11–36.

Anand, D. (2003). A contemporary story of 'diaspora': the tibetan version. *Diaspora* 12(2), 211–229.

Anand, D. (2004). A story to be told: IR, postcolonialism and the discourse of Tibetan (trans)national identity. In G. Chowdhry & S. Nair (eds), *Power, Postcolonialism and International Relations: Reading Race, Gender and Class*. London: Routledge, pp. 209–224.

Anand, D. (2006). The Tibet question and the West: issues of sovereignty, identity and representation. In J.T. Dreyer & B. Sautman (eds), *Contemporary Tibet: Politics, Development and Society in a Disputed Region*. London: M.E. Sharpe, pp. 285–304.

Anderson, B. (1991). *Imagined Communities: Reflections on the Origin and Spread of Nationalism*. London: Verso.

Anderson, B. (2006). Becoming and being hopeful: towards a theory of affect. *Environment and Planning D: Society and Space* 24, 733–752.

Anderson, B. (2010). Preemption, precaution, preparedness: Anticipatory action and future geographies. *Progress in Human Geography* 34(6), 777–798.

Anderson, J. (1996). The shifting stage of politics: new medieval and postmodern territorialities? *Environment and Planning D: Society and Space* 14(2), 133–153.

Anghie, A. (2002). Colonialism and the birth of international institutions: sovereignty, economy, and the mandate system of the League of Nations. *New York University Journal of International Law and Politics* 34(3), 513–633.

Appadurai, A. (1996). *Modernity at Large: Cultural Dimensions of Globalization*. Minneapolis, MN: University of Minnesota Press.

Appadurai, A. (2001). Deep democracy: urban governmentality and the horizon of politics. *Environment and Urbanization* 13(2) 23–43.

Arakeri, A.V. (1980). *Tibetans in India: The Uprooted People and their Cultural Transplantation*. New Delhi: Reliance Publishing House.

Ardley, J. (2002). *The Tibetan Independence Movement: Political, Religious and Gandhian Perspectives*. London: Routledge Curzon.

Ardley, J. (2003). Learning the art of democracy? Continuity and change in the Tibetan government-in-exile. *Contemporary South Asia* 12(3), 349–363.

Arendt, H. (1958). *The Human Condition*. Chicago, IL: University of Chicago Press.

Arendt, H. (1976). *The Origins of Totalitarianism*. San Diego, CA: Harvest.

Aretxaga, B. (2000). A fictional reality: paramilitary death squads and the construction of state terror in Spain. In J.A. Sluka (ed.), *Death Squad: The Anthropology of State Terror* Philadelphia: University of Pennsylvania Press, pp. 46–60.

Aretxaga, B. (2003). Maddening states. *Annual Review of Anthropology* 32, 393–410.

Armstrong, H.W. & Read, R. (2000). Comparing the economic performance of dependent territories and sovereign microstates. *Economic Development and Cultural Change* 48(1), 285–306.

Arpi, C. (2009). India Tibet relations 1947–1949: India begins to vacillate. Paper presented at the International Conference on Exploring Tibet's History and Culture. Delhi University, 19–21 November 2009.

Ashley, R.K. (1988). Untying the sovereign state: a double reading of the anarchy problematique. *Millennium: Journal of International Studies* 17(2), 227–262.

Bahcheli, T., Bartmann, B. & Srebrnik, H. (eds) (2004). *De facto States: The Quest for Sovereignty*. London: Routledge.

Barfield, T.J. (1989). *The Perilous Frontier: Nomadic Empires and China*. Cambridge, MA: Basil Blackwell.

Barker, R. (1990). *Political Legitimacy and the State*. Clarendon Press.

Barkin, J.S. & Cronin, B. (1994). The state and the nation: Changing norms and the rules of sovereignty in international relations. *International Organization* 48(1), 107–130.

Barnett, C. (2005). Ways of relating: hospitality and the acknowledgment of otherness. *Progress in Human Geography* 29, 5–21.

Barnett, R. (2001). "Violated specialness": Western political representations of Tibet. In T. Dodin & H. Räther (eds), *Imagining Tibet: Perceptions, Projections, and Fantasies*. Somerville, MA: Wisdom, pp. 269–316.

Barry, A., Osborne, T. & Rose, N. (1996). *Foucault and Political Reason: Liberalism, Neo-Liberalism and Rationalities of Government*. Chicago: Chicago University Press.

Bartelson, J. (1995). *A Genealogy of Sovereignty*. Cambridge: Cambridge University Press.

Bauman, Z. (1991). *Modernity and the Holocaust*. Ithaca, NY: Cornell University Press.

Bauman, Z. (2002). In the lowly nowherevilles of liquid modernity: Comments on and around Agier. *Ethnography* 3(3), 343–349.

Bayart, J.-F. (2007). *Global Subjects: A Political Critique of Globalization*. Cambridge: Polity Press.

Beckwith, C.I. (1993). *The Tibetan Empire in Central Asia : A History of the Struggle for Great Power among Tibetans, Turks, Arabs, and Chinese during the Early Middle Ages*. Princeton, NJ: Princeton University Press.

Beer, C. (2008). The spatial accommodation practice of the bureaucracy of the Commonwealth of Australia and the production of Canberra as national capital space: A dialectical and prosaic history. *Political Geography* 27, 40–56.

Beier, J.M. (ed.) (2010). *Indigenous Diplomacies*. Basingstoke, UK: Palgrave Macmillan.

Beker, M. & van Oordt, R. (1991). *The Palestinians in Lebanon: Contradictions of State-Formation in Exile*. Amsterdam: Middle East Associates.

Bell, C. (1924). *Tibet Past and Present*. Oxford: Clarendon Press.

Benhabib, S., Waldreon, J., Honig, B., Kymlicka, W. & Post, R. (2006). *Another Cosmopolitanism*. Oxford: Oxford University Press.

Bentz, A.-S. (2012). Symbol and power: the Dalai Lama as a charismatic leader. *Nations and Nationalism* 18(2), 287–305.

Bernstein, A. & Mertz, E. (2011). Introduction – bureaucracy: Ethnography of the state in everyday life. *Political and Legal Anthropology Review* 34(1), 6–10.

Bhabha, H. (1984). Of mimicry and man: the ambivalence of colonial discourse. *Discipleship: A Special Issue on Psychoanalysis* 28, 125–133.

Bickers, R. (2012). *The Scramble for China: Foreign Devils in the Qing Empire, 1832–1914*. London: Penguin.

Biersteker, T. & Weber, C. (1996). *State Sovereignty as Social Construct*. Cambridge: Cambridge University Press.

Billig, M. (1995). *Banal Nationalism*. London: Sage.

Blitz, B.K. (2011). Statelessness and environmental-induced displacement: Future scenarios of deterritorialisation, rescue and recovery examined. *Mobilities* 6(3), 433–450.

Blomley, N. (1994). Activism and the academy. *Environment and Planning D: Society and Space* 12(4), 383–385.

Bodh, A. (2011). 25-yr-old first Tibetan to be Indian citizen. *The Times of India*, 20 January. Available from http://timesofindia.indiatimes.com/india/25–yr–old–first–Tibetan–to–be–Indian–citizen/articleshow/7323090.cms?referral=PM (accessed 22 October 2012).

Boudreau, J.-A. (2001). Strategic territorialisation: the politics of Anglo-Montrealers. *Tijdschrift voor Economische en Sociale Geografie* 92(4), 405–419.

Bourdieu, P. (1994). Rethinking the state: Genesis and structure of the bureaucratic field. *Sociological Theory* 12(2), 1–18.

Bradatan, C. (2014). The wisdom of the exile. *New York Times*, 16 August. Available from http://opinionator.blogs.nytimes.com/2014/2008/2016/the–wisdom–of–the–exile/?_php=true&_type=blogs&_php=true&_type=blogs&_r=2011.

Brah, A. (1996). *Cartographies of Diaspora: Contesting Identities*. London: Routledge.

Brass, P. (1990). *The Politics of India since Independence*. Cambridge, UK: Cambridge University Press.

Bratsis, P. (2006). *Everyday Life and the State*. London: Paradigm Publishers.

Brenner, N., Jessop, B., Jones, M. & MacLeod, G. (eds) (2003). *State/Space: A Reader*. Oxford: Blackwell.

Brubaker, R. & Cooper, F. (2000). Beyond "Identity". *Theory and Society* 29(1), 1–47.

Burkett, M. (2011). The Nation *Ex-Situ*: On climate change, deterritorialized nationhood and the post-climate era. *Climate Law* 2, 345–374.

Butler, J. (1990). *Gender Trouble: Feminism and the Subversion of Identity*. London: Routledge.

Butler, J. (1992). *Bodies that Matter: On the Discursive Limits of Sex*. London: Routledge.

Buzan, B. (1991). *People, States and Fear: An Agenda for International Security Studies in the Post-Cold War Era*. New York: Harvester Wheatsheaf.

Cadman, L. (2010). How (not) to be governed: Foucault, critique, and the political. *Environment and Planning D: Society and Space* 28(3), 539–556.

Camilleri, J. & Falk, J. (1992). *The End of Sovereignty? The Politics of a Shrinking and Fragmenting World*. Aldershot: Edward Elgar.

Campbell, D. (1992). *Writing Security: United States Foreign Policy and the Politics of Identity*. Minneapolis, MN: University of Minnesota Press.

Candea, M. & da Col, G. (2012). The return to hospitality. *Journal of the Royal Anthropological Institute* 18(S1), S1–S19.

Caplan, J. & Torpey, J. (eds) (2001). *Documenting Individual Identity: The Development of State Practices in the Modern World*. Princeton, NJ: Princeton University Press.

Carrasco, P. (1959). *Land and Polity in Tibet*. Seattle: University of Washington Press.

Caspersen, N. (2012). *Unrecognized States*. Cambridge, UK: Polity Press.

Caspersen, N. (forthcoming). Degrees of legitimacy: Ensuring internal and external support in the absence of recognition. *Geoforum*

Cassidy, F. & Bish, R.L. (1989). *Indian Government: Its Meaning in Practice*. Lantzville, BC: Oolichan Books.

Castells, M. (1996). *The Rise of the Network Society*. Oxford: Blackwell.

Castoriadis, C. (1987). *The Imaginary Institution of Society*. Cambridge, UK: Polity Press.

Chakrabarty, D. (1999). Adda, Calcutta: dwelling in modernity. *Public Culture* 11(1), 109–145.

Chapman, F.S. (1938). *Lhasa: The Holy City*. London: Chatto & Windus.

Chatterjee, P. (2004). *The Politics of the Governed: Reflections on Popular Politics in Most of the World*. New York, Columbia University Press.

Chatterjee, P. (2005). The nation in heterogeneous time. *Futures* 37, 925–942.

Chatterji, J. (2007). 'Dispersal' and the failure of rehabilitation: Refugee camp-dwellers and squatters in West Bengal. *Modern Asian Studies* 41(5), 995–1032.

Chaturvedi, G. (2004). Indian visions. In D. Bernstorff & H. von Welck (eds), *Exile as Challenge: The Tibetan Diaspora*. New Delhi: Orient Longman, pp. 72–86.

Chellaney, B. (2011). *Water: Asia's New Battleground*. Washington, DC: Georgetown University Press.

Childs, G. (2008). *Tibetan Transitions: Historical and Contemporary Perspectives on Fertility, Family Planning, and Demographic Change*. Leiden: Brill.

Childs, G. & Barkin, G. (2006). Reproducing identity: Using images to promote postnatalism and sexual endogamy among Tibetan exiles in South Asia. *Visual Anthropology Review* 22(2), 34–52.

Chimni, B.S. (2003). Status of refugees in India: strategic ambiguity. In R. Samaddar (ed.), *Refugees and the State: Practices of Asylum and Care in India*. New Delhi: Sage, pp. 443–470.

Clapham, C. (1998). Degrees of statehood. *Review of International Studies* 24(2), 143–157.

Clayton, D. (2000). Governmentality. In R.J. Johnston, D. Gregory, G. Pratt, & M. Watts (eds), *The Dictionary of Human Geography*, 4th edn. Oxford: Blackwell, pp. 318–319.

Collier, S. (2008). Enacting catastrophe: preparedness, insurance, budgetary rationalisation. *Economy and Society* 37, 224–250.

Colson, E. (2003). Forced migration and the anthropological response. *Journal of Refugee Studies* 16(1), 1–18.

Constantinou, C.M. (1996). *On the Way to Diplomacy*. Minneapolis, MN: University of Minnesota Press.

Constantinou, C.M. (1998). Before the Summit: representations of sovereignty on the Himalayas. *Millennium: Journal of International Studies* 27(1), 23–53.

Cook, I. & Crang, P. (1995). *Doing Ethnographies: Concepts and Techniques in Modern Geography (CATMOG) 58*. Norwich: Institute of British Geographers' Quantitative Geography Study Group.

Conway, M. & Gotovitch, J. (2001). *Europe in Exile: European Exile Communities in Britain 1940–1945*. Oxford: Berghahn Books.

Corbridge, S., Williams, G., Srivistava, N. & Veron, R. (2005). *Seeing the State: Governance and Governmentality in India*. Cambridge, UK: Cambridge University Press.

Cornell, A. (2011). *Oppose and Propose: Lessons from Movement for a New Society* Oakland, CA: AK Press.

Corrigan, P. & Sayer, S. (1985). *The Great Arch: English State Formation as Cultural Revolution*. Oxford: Blackwell.

Cox, K.R. (1998). Spaces of dependence, spaces of engagement and the politics of scale, or: looking for local politics. *Political Geography* 17(1), 1–23.

Craggs, R. (2014). Hospitality in geopolitics and the making of Commonwealth international relations. *Geoforum* 52, 90–100.

Crang, P. (1994). "It's showtime": on the workplace geographies of display in a restaurant in Southeast England. *Environment and Planning D: Society and Space* 12(6), 675–704.

Crawford, J. (2006). *The Creation of States in International Law*. Oxford: Clarendon Press.

Cresswell, T. (1996). *In Place/Out of Place: Geography, Ideology and Transgression*. Minneapolis, MN: University of Minnesota Press.

CTRC (2003). *His Holiness the Dalai Lama's Central Tibetan Relief Committee: Building Sustainable Communities in Exile*. Dharamsala: Central Tibetan Relief Committee.

Dahl, R.A. (2000). *On Democracy*. New Haven: Yale University Press.

Dalai Lama (1963). Constitution of Tibet. New Delhi: Bureau of His Holiness The Dalai Lama.

Dalai Lama (1988). Strasbourg Proposal 1988: Address to Members of the European Parliament, 15 June. Available from http://www.dalailama.com/messages/tibet/strasbourg-proposal-1988 (accessed 14 September 2015).

Dalai Lama (1990). *Freedom in Exile: The Autobiography of the Dalai Lama*. London: Harper Collins.

Dalai Lama (1997). *My Land and My People: The Autobiography of His Holiness the Dalai Lama*. New York: Warner Books.

Darling, J. (2010). A city of sanctuary: the relational re-imagining of Sheffield's asylum politics. *Transactions of the Institute of British Geographers* 35(1), 125–140.

Das, V. & Poole, D. (eds) (2004). *Anthropology in the Margins of the State*. Santa Fe, NM: School of American Research Press.

Davidson, J. (2003). "Putting on a face": Sartre, Goffman and agoraphobic anxiety in social space. *Environment and Planning D: Society and Space* 21, 107–122.

Davies, A. (2012). Assemblage and social movements: Tibet Support Groups and the spatialities of political organisation. *Transactions of the Institute of British Geographers* 37(2), 273–286.

Davis, T.C. (2007). *Stages of Emergency: Cold War Nuclear Civil Defense*. Durham, NC: Duke University Press.

Dean, M. (1999). *Governmentality: Power and Rule in Modern Society*. London: Sage.

Dean, M. & Henman, P. (2004). Governing society today: editor's introduction. *Alternatives* 29, 483–494.

Department of Education CTA (2005). *Basic Education Policy for Tibetans in Exile*. Dharamsala: Department of Education, CTA.

Department of Finance (2005) *Paljor Bulletin* June. Dharamsala: Department of Finance, CTA.

Derrida, J. (1997) Politics and friendship: A discussion with Jacques Derrida. Interview at the Centre for Modern French Thought, University of Sussex, 1 December 1997. Available from http://www.livingphilosophy.org/Derrida-politics-friendship.htm (accessed 15 September 2015).

Derrida, J. (1999). *Adieu to Emmanuel Levinas*. Stanford, CA: Stanford University Press.

Desbiens, C., Mountz, A. & Walton-Roberts, M. (2004). Guest editorial. Introduction: reconceptualising the state from the margins of political geography. *Political Geography* 23(3), 241–243.

de Voe, D.M. (1987). Keeping refugee status: a Tibetan perspective. In S.M. Morgan & E. Colson (eds), *People in Upheaval*. New York: Centre for Migration Studies, pp. 54–65.

Dhondup, K. (1994). Dharamsala: Shangrila or Sarajevo? *Tibetan Review* 29(7), 14–20.

DIIR (1996). *Tibet: Proving Truth from Facts*. Dharamsala: DIIR, CTA.

DIIR (2005). *The Middle-Way Approach: A Framework for Resolving the Issue of Tibet*. Dharamsala: DIIR, CTA.

DIIR (2010). *Middle Way Policy and All Recent Related Documents*. Dharamsala: CTA.

Dodds, K., Kuus, M. & Sharp, J. (2013). Introduction: geopolitics and its critics. In K. Dodds, M. Kuus & J. Sharp (eds), *The Ashgate Research Companion to Critical Geopolitics*. Farnham: Ashgate, pp. 1–14.

Doty, R.L. (1996). Sovereignty and the nation: constructing the boundaries of national identity. In T.J. Biersteker & C. Weber (eds), *State Sovereignty as Social Construct*. Cambridge, UK: Cambridge University Press, pp. 121–147.

Dreyfus, G. (1995). Law, state and political ideology in Tibet. *Journal of the International Association of Buddhist Studies* 18(1), 117–138.

Dresser, N. (1997). On Sticking Out Your Tongue. *Los Angeles Times*, 8 November. Available from http://articles.latimes.com/1997/nov/08/local/me-51420 (accessed 15 September 2015).

Duara, P. (2003). *Sovereignty and Authenticity: Manchukuo and the East Asian Modern*. Lanham, MD: Rowman & Littlefield.

Duffield, M. (2010). Risk management and the fortified aid compound: every-day life in post-interventionary society. *Journal of Intervention and Statebuilding* 4, 453–474.

Dufoix, S. (2002). *Politiques d'exil. Hongrois, Polonais et Tchécoslovaques en France après 1945*. Paris: Presses Universitaires de France.

du Gay, P. (2000). *In Praise of Bureaucracy*. London: Sage.

Dulaney, A.G., Cusack, D.M. & van Walt van Praag, M. (1998). *The Case Concerning Tibet: Tibet's Sovereignty and the Tibetan People's Right to Self-determination*. New Delhi: TPPRC.

Duska, S.A. (2008). Harmony, ideology and dispute resolution: a legal ethnography of the Tibetan diaspora in India. PhD thesis, University of British Columbia, Vancouver.

Easton, D. (1975). A re-assessment of the concept of political support. *British Journal of Political Science* 5(4), 435–457.

Edensor, T. (ed.) (2010). *Geographies of Rhythm: Nature, Place, Mobilities and Bodies*. Aldershot: Ashgate.

Edin, M. (1992). Transition to democracy in exile: A study of the Tibetan Government's strategy for self-determination. MPhil thesis, University of Uppsala, Sutteanum.

Edkins, J., Persram, N. & Pin-Fat, V. (eds) (1999). *Sovereignty and Subjectivity*. Boulder, CO: Lynne Riener.

Elden, S. (2005). Missing the point: globalization, deterritorialization and the space of the world. *Transactions of the Institute of British Geographers* 30(1), 8–19.

Elden, S. (2006). Contingent sovereignty, territorial integrity and the sanctity of borders. *The SAIS Review of International Affairs* 26(1), 11–24.

Elden, S. (2007). Governmentality, calculation, territory. *Environment and Planning D: Society and Space* 25(3), 562–580.

Elverskog, J. (2006). *Our Great Qing: The Mongols, Buddhism, And the State in Late Imperial China*. Honolulu: University of Hawai'i Press.

Erman, E. & Uhlin, A. (2010). *Legitimacy Beyond the State?: Re-examining the Democratic Credentials of Transnational Actors*. Basingstoke: Palgrave.

Farah, R. (2009). Refugee camps in the Palestinian and Sahrawi National Liberation Movements. *Journal of Palestine Studies* 38(2), 76–93.

Feldman, I. (2008). *Governing Gaza: Bureaucracy, Authority and the Work of Rule (1917–67)*. Durham, NC: Duke University Press.

Ferguson, J. (1990). *The Anti-Politics Machine: "Development," Depoliticization, and Bureaucratic Power in Lesotho*. Cambridge, UK: Cambridge University Press.

Ferguson, J. & Gupta, A. (2002). Spatializing states: Toward an ethnography of neoliberal governmentality. *American Ethnologist* 29(4), 981–1002.

Ferguson, Y.H., & Mansbach, R.W. (1996). *Polities: Authority, Identities, and Change*. Columbia, SC: University of South Carolina Press.

Flint, C. (2009). State. In D. Gregory, R. Johnston, G. Pratt & S. Whatmore (eds), *The Dictionary of Human Geography*, 5th edn. Oxford: Blackwell, pp. 722–724.

Foley, G., Schaap, A. & Howell, E. (eds) (2013). *The Aboriginal Tent Embassy: Sovereignty, Black Power, Land Rights and the State*. London: Routledge.

Forsberg, T. (1996). Beyond sovereignty, within territoriality: mapping the space of late-modern (geo)politics. *Cooperation and Conflict* 31(4), 355–386.

Fortier, A.-M. (1999). Re-membering places and the performance of belonging(s). *Theory, Culture and Society* 16(2), 41–64.

Foucault, M. (1977). *Discipline and Punish: The Birth of the Prison* (A. Sheridan, trans.). London: Penguin Books.

Foucault, M. (1991). Governmentality. In G. Burchell, C. Gordon & P. Miller (eds), *The Foucault Effect: Studies in Governmentality*. Chicago: University of Chicago Press, pp. 87–104.

Foucault, M. (2004). *Naissance de la biopolitique. Cours au Collège de France, 1978–1979*. Paris: Seuil/Gallimard.

Foucault, M. (2007). *Security, Territory, Population: Lectures at the Collége de France 1977–78*. New York: Picador.

Frechette, A. (1997). Statelessness and power: transformational entitlements among Tibetan exiles in Kathmandu, Nepal. PhD thesis, Harvard University.

Frechette, A. (2006). Constructing the state in the Tibetan diaspora. In K. Boyd & T.-W. Ngo (eds), *State Making in Asia*. London: Routledge, pp. 127–149.

French, R.R. (1991). The new Snow Lion: The Tibetan Government-in-exile in India. In Y. Shain (ed.), *Governments-in-exile in Contemporary World Politics*. London: Routledge, pp. 188–201.

French, R.R. (1995). *The Golden Yoke: The Legal Cosmology of Buddhist Tibet*. Ithaca, NY: Cornell University Press.

Fuchs, A. & Klann, N.-H. (2010). Paying a Visit: The Dalai Lama Effect on International Trade. Center for European Governance and Economic Development, Research Paper No. 113.

Fürer-Haimendorf, C. (1990). *The Renaissance of the Tibetan Civilization*. London: Synergetic Press.

Garfield, J.L. (2002). *Empty Words: Buddhist Philosophy and Cross-cultural Interpretation*. Oxford: Oxford University Press.

Garratt, K. (1997). Tibetan refugees, asylum seekers, returnees and the Refugees' Convention – predicaments, problems and prospects. *Tibet Journal* 22(3), 18–56.

Geertz, C. (1980). *Negara: The Theatre State in Nineteenth-Century Bali*. Princeton: Princeton University Press.

George, J. (1994). *Discourses of Global Politics: A Critical (Re)Introduction to International Relations*. Boulder, CO: Lynne Rienner.

Gibson-Graham, J.K. (1996). *The End of Capitalism (As We Knew It): A Feminist Critique of Political Economy*. Oxford: Blackwell.

Giddens, A. (1985). *The Nation-State and Violence*. Cambridge, UK: Polity Press.

Gill, N. (2010). New state-theoretic approaches to asylum and refugee geographies. *Progress in Human Geography* 34(5), 626–645.

Goddeeris, I. (2007). The temptation of legitimacy: exile politics from a comparative perspective. *Contemporary European History* 16(3), 395–405.

Goffman, E. (1959). *The Presentation of Self in Everyday Life*. New York: Doubleday Anchor.

Goldman, M. (2006). *Imperial Nature: The World Bank and Struggle for Social Justice in the Age of Globalization*. New Haven: Yale University Press.

Goldstein, M.C. (1971a). The balance between centralization and decentralization in the traditional Tibetan political system: an essay on the nature of Tibetan political macrostructure. *Central Asiatic Journal* 15, 170–182.

Goldstein, M.C. (1971b). Serfdom and mobility: an examination of the institution of "human lease" in traditional Tibetan society. *The Journal of Asian Studies* 30(3), 521–534.

Goldstein, M.C. (1975). Tibetan refugees in South India: A new face to the Indo-Tibetan interface. *The Tibet Society Bulletin* 9, 12–29.

Goldstein, M.C. (1978). Ethnogenesis and resource competition among Tibetan refugees in South India: A new face to the Indo-Tibetan interface. In J.F. Fisher (ed.), *Himalayan Anthropology: The Indo-Tibetan Interface*. The Hague: Mouton Publishers, pp. 395–420.

Goldstein, M.C. (1989). *A History of Modern Tibet, 1913–1951, Volume 1 the Demise of the Lamaist state*. Berkeley, CA: University of California Press.

Gordillo, G. (2006). The crucible of citizenship: ID-paper fetishism in the Argentinian Chaco. *American Ethnologist* 33(2), 162–176.

Gottlieb, G. (1993). *Nation Against State: A New Approach to Ethnic Conflicts and the Decline of Sovereignty*. New York: Council on Foreign Relations Press.

Gould, A. (1993). *Capitalist Welfare Systems*. New York: Longman.

Gregson, N. & Rose, G. (2000). Taking Butler elsewhere: performativities, spatialities and subjectivities. *Environment and Planning D: Society and Space* 18(4), 433–452.

Grimmelmann, J. (2012). Sealand, HavenCo, and the rule of law. *University of Illinois Law Review* (2), 405–484.

Grunfeld, A.T. (1987). *The Making of Modern Tibet*. London: Zed Books.

Gupta, A. (1995). Blurred boundaries: the discourse of corruption, the culture of politics, and the imagined state. *American Ethnologist* 22(2), 375–402.

Gupta, A. (2007). Imagining nations. In D. Nugent & J. Vincent (eds), *A Companion to the Anthropology of Politics*. Oxford: Blackwell, pp. 267–281.

Gupta, A. (2012). *Red Tape: Bureaucracy, Structural Violence and Poverty in India*. Durham, NC: Duke University Press.

Gupta, A. & Ferguson, J. (1992). Beyond "culture": space, identity, and the politics of difference. *Cultural Anthropology* 7(1), 6–23.

Gyalpo, D. (2004a). 44 years of democracy in exile. *Tibetan Bulletin* 8(5), 24–26.

Gyalpo, D. (2004b). Exile Tibetans in Statistics: Book Review of 'Tibetan Community in Exile: Demographic and Socio–Economic Issues 1998–2001, Planning Commission, CTA. Phayul.com, 13 October. Available at www.phayul.com/news/article.aspx?id=7948<=4&c=1 (accessed 14 September 2015).

Hampton, T. (2009). *Fictions of Embassy: Literature and Diplomacy in Early Modern Europe.* New York: Cornell University Press.

Hanafi, S. (2008). Palestinian refugee camps in Lebanon: Laboratories of state-in-the-making, discipline and Islamist radicalism. In R. Lentin (ed.), *Thinking Palestine.* London: Zed Books, pp. 82–100.

Hannah, M. (2000). *Governmentality and the Mastery of Territory in Nineteenth-Century America.* Cambridge, UK: Cambridge University Press.

Hannah, M. (2009). Calculable territory and the West German census boycott movements of the 1980s. *Political Geography* 28, 66–75.

Hansen, P. (2003). Why is there no subaltern studies in Tibet? *The Tibet Journal* XXVIII(4), 7–22.

Hansen, T.B. (2001). Governance and state mythologies in Mumbai. In T.B. Hansen & F. Stepputat (eds), *States of Imagination: Ethnographic Explorations of the Postcolonial State.* Durham, NC: Duke University Press, pp. 221–256.

Hansen, T.B. (2005). Sovereigns beyond the state: on legality and authority in urban India. In T.B. Hansen & F. Stepputat (eds), *Sovereign Bodies: Citizens, Migrants and States in the Postcolonial World.* Princeton, NJ: Princeton University Press, pp. 169–191.

Hansen, T.B. & Stepputat, F. (2001). Introduction: states of imagination. In T.B. Hansen & F. Stepputat (eds), *States of Imagination: Ethnographic Explorations of the Postcolonial State.* Durham, NC: Duke University Press, pp. 1–40.

Hansen, T.B. & Stepputat, F. (eds) (2005). *Sovereign Bodies: Citizens, Migrants and States in the Postcolonial World.* Princeton, NJ: Princeton University Press.

Harris, C. (1997). Struggling with Shangri-la: A Tibetan artist in exile. In F. Korom (ed.), *Constructing Tibetan Culture: Contemporary Perspectives.* Quebec: World Heritage Press, pp. 160–177.

Held, D. (2006). *Models of Democracy,* 3rd edn. Cambridge, UK: Polity Press.

Herbert, S. (2000). For ethnography. *Progress in Human Geography* 24(4), 550–568.

Hertzfeld, M. (2012). Afterword: reciprocating the hospitality of these pages. *Journal of the Royal Anthropological Institute* 18(S1), S210–S217.

Herzfeld, M. (1992). *The Social Production of Indifference: Exploring the Symbolic Roots of Western Bureaucracy.* Chicago: University of Chicago Press.

Hess, J.M. (2006). Statelessness and the state: Tibetans, citizenship, and nationalist activism in a transnational world. *International Migration* 44(1), 79–101.

Hess, J.M. (2009). *Immigrant Ambassadors: Citizenship and Belonging in the Tibetan Diaspora.* Stanford, CA: Stanford University Press.

Hewitt, K. (1983). Place annihilation: Area bombing and the fate of urban places. *Annals of the Association of American Geographers* 73(2), 257–284.

Heyman, J. (1995). Putting power in the anthropology of bureaucracy: the Immigration and Naturalization Service at the Mexico-United States border. *Current Anthropology* 36(2), 261–287.

Higate, P. & Henry, M. (2009). *Insecure Spaces: Peacekeeping, Power and Performance in Haiti, Kosovo and Liberia.* London: Zed Books.

Hoag, C. (2011). Assembling partial perspectives: thoughts on the anthropology of bureaucracy. *Political and Legal Anthropology Review* 34(1), 81–94.

Hobsbawm, E. & Ranger, T. (1983). *The Invention of Tradition*. Cambridge, UK: Cambridge University Press.

Hocking, B. (1999). Patrolling the 'frontier': globalization, localization and the 'actorness' of non-central governments. In F. Aldecoa & M. Keating (eds), *Paradiplomacy in Action: The Foreign Relations of Subnational Governments*. Portland, OR: Frank Cass, pp. 17–39.

Hodges, S. (2004). Governmentality, population and reproductive family in modern India. *Economic and Political Weekly*, 13 March, 1157–1163.

Home Office (2003). *China Country Report, October 2003*. London: Immigration and Nationality Directorate, Home Office UK.

hooks, b. (1990). Marginality as a site of resistance. In R. Ferguson (ed.), *Out There: Marginalization and Contemporary Cultures*. Cambridge, MA: MIT Press, pp. 341–343.

Horowitz, D.L. (1998). Self-determination: politics, philosophy and law. In M. Moore (ed.), *National Self-Determination and Secession*. Oxford: Oxford University Press, pp. 181–214.

Horowitz, L.S. (2009). Environmental violence and crises of legitimacy in New Caledonia. *Political Geography* 28, 248–258.

Houston, S. & Wright, R. (2003). Making and remaking Tibetan diasporic identities. *Social and Cultural Geography* 4(2), 217–231.

Howitt, R. (1998). Scale as relation: musical metaphors of geographical scale. *Area* 30(1), 49–58.

Howitt, R. (2003). Scale. In J. Agnew, K. Mitchell & G. Toal (eds), *A Companion to Political Geography*. Oxford: Blackwell, pp. 138–157.

Huber, T. (2001). Shangri-la in exile: representations of Tibetan identity and transnational culture. In T. Dodin & H. Räther (eds), *Imagining Tibet: Perceptions, Projections, and Fantasies*. Somerville, MA: Wisdom, pp. 357–371.

Human Rights Watch (2010). *"I saw it with my own eyes": Abuses by Chinese security forces in Tibet, 2008–2010*. New York: Human Rights Watch

Human Rights Watch (2013). *"They say we should be grateful": Mass rehousing and relocation programs in Tibetan areas of China*. New York: Human Rights Watch.

Immigration and Refugee Board of Canada (1998). CHN30410.E:The requirements a Tibetan-in-exile must meet to obtain a Green Book. Available from UNHCR refworld: http://www.refworld.org/docid/3ae6ac8d64.html (accessed 14 September 2015).

International Campaign for Tibet (2012). Refugee Report: Dangerous Crossing – 2011 Update. Washington DC: International Campaign for Tibet.

Isin, E.F. (2008). Theorizing acts of citizenship. In E.F. Isin & G.M. Nielsen (eds), *Acts of Citizenship*. London: Zed Books, pp. 15–43.

Isin, E.F. & Turner, B.S. (2007). Investigating citizenship: an agenda for citizenship studies. *Citizenship Studies* 11(1), 5–17.

Jackson, R. (1990). *Quasi-states: Sovereignty, International Relations and the Third World*. Cambridge, UK: Cambridge University Press.

Jeffrey, A. (2006). Building state capacity in post-conflict Bosnia and Herzegovina: The case of Brčko District. *Political Geography* 25(2), 203–227.

Jeffrey, A. (2013). *The Improvised State: Sovereignty, Performance and Agency in Dayton Bosnia*. Oxford: Wiley-Blackwell.

Jeffrey, C. (2008). Guest editorial: waiting. *Environment and Planning D: Society and Space* 26(6), 954–958.

Jeffrey, C. (2010). *Timepass: Youth, Class and the Politics of Waiting in India*. Stanford, CA: Stanford University Press.

Jeffrey, C., Jeffrey, P. & Jeffrey, R. (2008). *Degrees Without Freedom? Education, Masculinities and Unemployment in North India*. Stanford, CA: Stanford University Press.

Jessop, B. (1990). *State Theory: Putting Capitalist States in Their Place*. Cambridge, UK: Polity Press.

Jessop, B. (1997). A neo-Gramscian approach to the regulation of urban regimes: accumulation strategies, hegemonic projects, and governance. In M. Lauria (ed.), *Reconstructing Urban Regime Theory: Regulating Urban Politics in a Global Economy*. Thousand Oaks, CA: Sage, pp. 51–73.

Jessop, B. (2007). From micro-powers to governmentality: Foucault's work on statehood, state formation, statecraft and state power. *Political Geography* 26(1), 34–40.

Jessop, B., Brenner, N. & Jones, M. (2008). Theorizing sociospatial relations. *Environment and Planning D: Society and Space* 26(3), 389–401.

Jerryson, M. & Juergensmeyer, M. (eds) (2010). *Buddhist Warfare*. Oxford: Oxford University Press

Jones, R. (2009). Categories, borders and boundaries. *Progress in Human Geography* 33(2), 174–189.

Jones, R. (2007). *People/States/Territories*. Oxford: Wiley-Blackwell.

Jones, R. (2012). State encounters. *Environment and Planning D: Society and Space* 30(5), 805–821.

Kantorowicz, E. (1957/1981). *The King's Two Bodies, a Study in Medieval Political Theology*. Princeton, NJ: Princeton University Press.

Kashag (2012). Kalon Tripa accepts resignations of special envoy Lodi G. Gyari and envoy Kelsang Gyaltsen. Central Tibetan Administration press release, 3 June. Available from http://tibet.net/2012/06/03/kalon-tripa-accepts-resignations-of-special-envoy-lodi-g-gyari-and-envoy-kelsang-gyaltsen/ (accessed 14 September 2015).

Kearns, G. (2008). Progressive geopolitics. *Geography Compass* 2(5): 1599–1620.

Keating, M. (2001). *Plurinational Democracy: Stateless Nations in a Post-Sovereignty Era*. Oxford: Oxford University Press.

Kharat, R. (2003). *Tibetan Refugees in India*. New Delhi: Kaveri Books.

Kingston, P. (2004). Introduction. States-within-states: historical and theoretical perspectives. In P. Kingston & I.S. Spears (eds), *States-Within-States: Incipient Political Entities in the Post Cold War Era*. New York: Palgrave Macmillan, pp. 1–13.

Klieger, P.C. (1992). *Tibetan Nationalism: The Role of Patronage in the Accomplishment of a National Identity*. Berkeley, CA: Folklore Institute.

Klieger, P.C. (2002). Introduction: The quest for understanding the modern Tibetan self. In P.C. Klieger (ed.), *Tibet, Self, and the Tibetan Diaspora: Voices of Difference*. Leiden: Brill, pp. 1–9.

Knaus, J.K. (2000). *Orphans of the Cold War: America and the Tibetan Struggle for Survival*. New York: PublicAffairs.

Kolås, A. (1996). Tibetan nationalism: the politics of religion. *Journal of Peace Research* 33(1), 51–66.

Kolossov, V.A. & O'Loughlin, J. (1998). Pseudo-states as harbingers of a new geopolitics: the example of the Trans-Dniester Moldovan Republic (TMR). *Geopolitics* 3(1), 151–176.

Kondapalli, S. & Mifune, E. (eds) (2010). *China and its Neighbours.* New Delhi: Pentagon Press.

Korom, F. (ed.) (1997). *Constructing Tibetan Culture: Contemporary Perspectives.* Quebec: World Heritage Press.

Krader, L. (1963). *Social organisation of the Mongol-Turkic pastoral nomads.* The Hague: Mouton.

Krasner, S.D. (1999). *Sovereignty: Organised Hypocrisy.* Princeton, NJ: Princeton University Press.

Krishna, S. (1992). Oppressive pasts and desired futures: re-imagining India. *Futures* 24(9), 858–866.

Kuleshov, N.S. (1996). *Russia's Tibet File.* Dharamsala: Library of Tibetan Works and Archives (LTWA).

Kuus, M. (2008). Professionals of geopolitics: agency in international politics. *Geography Compass* 2(6), 2062–2079.

Kuus, M. (2011). Bureaucracy and place: expertise in the European Quarter. *Global Networks* 11(4), 421–439.

Kuus, M. (2013). Foreign policy and ethnography: a sceptical intervention. *Geopolitics* 18(1), 115–131.

Kuus, M. (2014). *Geopolitics and Expertise: Knowledge and Authority in European Diplomacy.* Oxford: Wiley-Blackwell.

Kuzmin, S.L. (2011). *Hidden Tibet: History of Independence and Occupation.* Dharamsala: Library of Tibetan Works and Archives (LTWA).

Lafitte, G. (1999). Tibetan futures: imagining collective destinies. *Futures* 31(2), 155–169.

Latour, B. (1987). *Science in Action: How to Follow Scientists and Engineers through Society.* Cambridge, MA: Harvard University Press.

Lattimore, O. (1940/1949). *Inner Asian Frontiers of China.* New York: American Geographical Society, Research Series No. 21, 2nd edn.

Lau, T. (2009). Tibetan fears and Indian foes: fears of cultural extinction and antagonism as discursive strategy. *Explorations in Anthropology* 9(1), 81–90.

Lecours, A. (2002). Paradiplomacy: Reflections on the foreign policy and international relations of regions. *International Negotiation* 7(1), 91–114.

Lefebvre, H. (2004). *Rhythmanalysis: Space, Time and Everyday Life* (S. Elden & G. Moore, trans.). London: Continuum.

Legg, S. (2005). Foucault's population geographies: classifications, biopolitics and governmental spaces. *Population, Space and Place* 11(3), 137–156.

Lei, H. (2011). Foreign Ministry spokesperson Hong Lei's regular press conference on September 26, 2011. Available from http://uy.china-embassy.org/eng/fyrth/t863132.htm (accessed 16 September 2015).

Lemay-Hébert, N. (2009). Statebuilding without nation-building? Legitimacy, state failure and the limits of the institutionalist approach. *Journal of Intervention and Statebuilding* 3(1), 21–45.

Ley, D. (2000). Geography of spectacle. In R.J. Johnston, D. Gregory, G. Pratt & M. Watts (eds), *The Dictionary of Human Geography*, 4th edn. Oxford: Blackwell, p. 872.

Lipset, S.M. (1984). Social conflict, legitimacy, and democracy. In W. Connolly (ed.), *Legitimacy and the State*. New York, New York University Press, pp. 88–103.

Lopez, D.S. (1998). *Prisoners of Shangri-la: Tibetan Buddhism and the West*. Chicago: University of Chicago Press.

Luke, T.W. (1996). Governmentality and contragovernmentality: rethinking sovereignty and territoriality after the Cold War. *Political Geography* 15(6/7), 491–507.

Lynch, M. (2012). *The Arab Uprising: The Unfinished Revolutions of the New Middle East*. New York, PublicAffairs.

MacLeod, R. (1988). Introduction. In R. MacLeod (ed.), *Government and Expertise: Specialists, Administrators and Professionals, 1860–1919*. Cambridge, UK: Cambridge University Press, pp. 1–24.

Mains, D. (2007). Neoliberal times: progress, boredom and shame among young men in urban Ethiopia. *American Ethnologist* 34(4), 659–673.

Malkki, L. (1992). National Geographic: the rooting of peoples and the territorialization of national identity among scholars and refugees. *Cultural Anthropology* 7(1), 24–44.

Malkki, L. (1995). *Purity and Exile: Violence, Memory, and National Cosmology among Hutu Refugees in Tanzania*. Chicago: Chicago University Press.

Mandaville, P.G. (1999). Territory and translocality: discrepant idioms of political identity. *Millennium: Journal of International Studies* 28(3), 653–673.

Mann, M. (1984). The autonomous power of the state: its origins, mechanisms and results. *Archives Européennes de Sociologie* 25, 185–213.

Marcus, G.E. (1995). Ethnography in/of the world system: the emergence of multi-sited ethnography. *Annual Review of Anthropology* 24, 95–117.

Marston, S.A. (2004). Space, culture, state: uneven developments in political geography. *Political Geography* 23(1), 1–16.

Massey, D. (2005). *For Space*. London: Sage.

Mayall, J. & de Oliveira, R.S. (eds) (2011). *The New Protectorates: International Tutelage and the Making of Liberal States*. London: C. Hurst & Co.

McAdam, J. (2010). 'Disappearing states', statelessness and the boundaries of international law. *UNSW Law Research Paper No. 2010-2*.

McConnell, F. (2009a). De facto, displaced, tacit: the sovereign articulations of the Tibetan government-in-exile. *Political Geography* 28(6), 343–352.

McConnell, F. (2009b). Governments-in-exile: statehood, statelessness and the reconfiguration of territory and sovereignty. *Geography Compass* 3(5), 1912–1919.

McConnell, F. (2011). A state within a state? Exploring relations between the Indian state and the Tibetan community and government-in-exile. *Contemporary South Asia* 18(3), 297–313.

McConnell, F. (2012). Governmentality to practise the state? Constructing a Tibetan population in exile. *Environment and Planning D: Society and Space* 30(1), 78–95.

McConnell, F. (2013a). Citizens and refugees: constructing and negotiating Tibetan identities in exile. *Annals of the Association of American Geographers* 103(4), 967–983.

McConnell, F. (2013b). The geopolitics of Buddhist reincarnation: contested futures of Tibetan leadership. *Area* 45(2), 162–169.

McConnell, F. (2013c). Tibetan democracy in exile: the 'uniqueness' and limitations of democratic procedures in a territory-less polity. In V. Arora & N. Jayaram (eds), *Routeing Democracy in the Himalayas: Experiments and Experiences*. New Delhi: Routledge, pp. 204–231.

McConnell, F. (2014). Contextualising and contesting peace: geographies of Tibetan satyagraha. In N. Megoran, F. McConnell & P. Williams (eds), *Geographies of Peace*. London: I B Tauris, pp. 131–150

McConnell, F. (2015). Reconfiguring diaspora identities and homeland connections: the Tibetan 'Lhakar' movement. In M.E. Christou & A. Christou (eds), *Dismantling Diasporas: Rethinking the Geographies of Diasporic Identity, Connection and Development*. Farnham: Ashgate, pp. 99–111.

McConnell, F., Moreau, T. & Dittmer, J. (2012). Mimicking state diplomacy: The legitimizing strategies of unofficial diplomacies. *Geoforum* 43(4), 804–814.

McCorquodale, R. & Orosz, N. (eds) (1994). *Tibet: The Position in International Law, Report of the Conference of International Lawyers on Issues Relating to Self-Determination and Independence for Tibet, London 6–10 January 1993*. Stuttgart: Hansjörg Mayer.

McDowell, L. & Court, G. (1994). Performing work: bodily representations in merchant banks. *Environment and Planning D: Society and Space* 12, 727–750.

McGranahan, C. (2007). Empire out of bounds: Tibet in the era of decolonization. In A. Stoler, C. McGranahan & P. Perdue (eds), *Imperial Formations*. Santa Fe, NM: SAR Press, pp. 187–227.

McGranahan, C. (2010). *Arrested Histories: Tibet, the CIA and Memories of a Forgotten War*. Durham, NC: Duke University Press.

McGukin, E.A. (1997). Postcards from Shangri-la: tourism, Tibetan refugees and the politics of cultural production. PhD thesis, City University New York, New York.

McKay, A. (1997). *Tibet and the British Raj: The Frontier Cadre, 1904–1947*. Richmond, UK: Curzon Press.

McLagan, M. (1996). Mobilising for Tibet: transnational politics and diaspora culture in the post Cold War era. PhD thesis, New York University, New York.

Megoran, N. (2006). For ethnography in political geography: experiencing and re-imagining Ferghana Valley boundary closures. *Political Geography* 25(6), 622–640.

Mehrotra, L.L. (2000). *India's Tibet Policy: An Appraisal and Options*, 3rd edn. New Delhi: TPPRC.

Migdal, J. (1988). *Strong States and Weak Societies: State–Society Relations and State Capabilities in the Third World*. Princeton: Princeton University Press.

Miller, P. & Rose, M. (1990). Governing economic life. *Economy and Society* 19(1), 1–31.

Mills, M. (2003a). *Identity, Ritual and State in Tibetan Buddhism: The Foundations of Authority in Gelukpa Monasticism*. London: Routledge.

Mills, M. (2003b). This turbulent priest: contesting religious rights and the state in the Tibetan Shugden controversy. In R. Wilson & J.P. Mitchell (eds), *Human Rights in Global Perspective: Anthropological Studies of Rights, Claims and Entitlements*. London: Routledge, pp. 54–70.

Mills, M. (2014). Who belongs to Tibet? Governmental narratives of state in the Ganden Podrang. In G. Toffin & J. Pfaff-Czarnecka (eds), *Facing Globalization in the Himalayas: Belonging and the Politics of the Self*. New Delhi: SAGE, pp. 397–418.

Minca, C. (2005). The return of the camp. *Progress in Human Geography* 29, 405–412.

Ministry of Foreign Affairs (1959). *Notes, Memoranda and Letters Exchanged and Agreements Signed by the Government of India and China, Vol. 2.* Delhi: Ministry of Foreign Affairs.

Mitchell, T. (1991). The limits of the state: beyond statist approaches and their critics. *American Political Science Review* 85(1), 77–96.

Mitchell, T. (2002). *Rule of Experts: Egypt, Techno-politics, Modernity.* Berkeley: University of California Press.

Miyazaki, H. (2004). *The Method of Hope: Anthropology, Philosophy, and Fijian Knowledge.* Stanford, CA: Stanford University Press.

Moncrieffe, J. & Eyben, R. (2007). *The Power of Labelling: How People are Categorized and Why it Matters.* London: Earthscan.

Montevideo Convention on Rights and Duties of States (1933). 165 L.N.T.S. 19. 26 December 1933.

Moshe, S. (1996). *The Palestinian Entity, 1959–1974 : Arab Politics and the PLO.* London: Cass.

Mountz, A. (2010). *Seeking Asylum: Human Smuggling and Bureaucracy at the Border.* Minneapolis: University of Minnesota Press.

Mountz, A. (2011). Where asylum-seekers wait: feminist counter-topographies of sites between states. *Gender, Place and Culture* 18, 381–399.

Moynihan, M. (1997). Tibetan refugees in India and Nepal, testimony by Maura Moynihan, consultant to Refugees International, before the Senate Foreign Relations Committee Hearing on Tibet. Available from http://www.radioradicale.it/exagora/tibetan-refugees-in-india-and-nepal-testimony-by-maura-moynihan-ict (accessed 16 September 2015).

Mundy, J.A. (2007). Performing the nation, prefiguring the state: the Western Saharan refugees, thirty years later. *Journal of Modern African Studies* 45(2), 275–297.

Murdoch, J. & Ward, N. (1997). Governmentality and territoriality: the statistical manufacture of Britain's 'National Farm'. *Political Geography* 16(4), 307–324.

Murphy, A.B. (1996). The sovereign state system as political-territorial ideal: Historical and contemporary considerations. In T.J. Biersteker & C. Weber (eds), *State Sovereignty as Social Construct.* Cambridge, UK: Cambridge University Press, pp. 81–120.

Murphy, S. (1999). Democratic legitimacy and the recognition of states and governments. *International and Comparative Law Quarterly* 48(3), 545–581.

Navaro-Yashin, Y. (2002). *Faces of the State: Secularism and Public Life in Turkey.* Princeton: Princeton University Press.

Navaro-Yashin, Y. (2003). 'Life is dead here': Sensing the political in 'no man's land'. *Anthropological Theory* 3(1), 107–125.

Navaro-Yashin, Y. (2005). Confinement and imagination: sovereignty and subjectivity in a quasi-state. In T.B. Hansen & F. Stepputat (eds), *Sovereign Bodies: Citizens, Migrants and States in the Postcolonial World* Princeton: Princeton University Press, pp. 103–119.

Navaro-Yashin, Y. (2006). Affect in the civil service: a study of a modern state-system. *Postcolonial Studies* 9(3), 281–294.

Navaro-Yashin, Y. (2012). *The Make-Believe Space: Affective Geography in a Postwar Polity.* Durham, NC: Duke University Press.

Neumann, I.B. (2012). *At Home with the Diplomats: Inside a European Foreign Ministry.* Ithaca, NY: Cornell University Press.

Nevins, J. (2002). *Operation Gatekeeper: The Rise of the "Illegal Alien" and the Making of the U.S.-Mexico Boundary*. London: Routledge.

Nielsen, M. (2007). Filling in the blanks: the potency of fragmented imageries of the state. *Review of African Political Economy* 34(114), 695–708.

Norbu, D. (1976). Editorial: towards sectarian harmony and national unity. *Tibetan Review* 11(9), 3–4.

Norbu, D. (1992). 'Otherness' and the modern Tibetan ientity. *Himal* May/June, 10–11.

Norbu, D. (2001). Refugees from Tibet: structural causes of successful settlements. *The Tibet Journal* 26(2), 3–25.

Norbu, D. (2004). The settlements: participation and integration. In D. Bernstorff & H. Von Welck (eds), *Exile as Challenge: The Tibetan Diaspora*. New Delhi: Orient Longman, pp. 186–212.

Norbu, J. (2004). Opening of the political eye: Tibet's long search for democracy. In J. Norbu (ed.), *Shadow Tibet*. New Delhi: Bluejay Books, pp. 11–25.

Norbu, J. (2007). Rangzen! *Combat Law: The Human Rights and Law Bimonthly* 6 (September–October), 28–35.

Nordholt, H. (1996). *The Spell of Power: A History of Balinese Politics, 1650–1940*. Leiden: KITLV Press.

Nowak, M. (1978). Liminal 'self', ambiguous 'power': the genesis of the 'Rangzen' metaphor among Tibetan youth in India. PhD thesis, University of Washington.

Nugent, D. (2007). Governing states. In D. Nugent & J. Vincent (eds), *A Companion to the Anthropology of Politics*. Oxford: Blackwell, pp. 198–215.

Nuijten, M. (2003). *Power, Community and the State: The Political Anthropology of Organisation in Mexico*. Sterling, VA: Pluto Press.

Nye, J.S. (2004). *Soft Power: The Means to Success in World Politics*. New York: PublicAffairs.

O'Brien, R. (1992). *Global Financial Integration: The End of Geography*. London: Chatham House Papers.

Ó Tuathail, G. & Luke, T. (1994). Present at the (dis)integration: deterritorialisation and reterritorialisation in the new wor(l)d order. *Annals of the Association of American Geographers* 84(3), 381–398.

Oberoi, P. (2006). *Exile and Belonging: Refugees and State policy in South Asia*. New Delhi: Oxford University Press.

Office of The Dalai Lama. (1969). *Tibetans in Exile 1959–1969: A Report on Ten Years of Rehabilitation in India*. New Delhi: Bureau of The Dalai Lama.

Ohmae, K. (1996). *End of the Nation State: The Rise of Regional Economies*. London: Harper Collins.

Ong, A. (1999). *Flexible Citizenship: The Cultural Logic of Transnationality*. Durham, NC: Duke University Press.

Pack, J. (2013). *The 2011 Libyan Uprisings and the Struggle for the Post-Qadhafi Future*. New York: Palgrave Macmillan.

Paljor Bulletin (2005). Dharamsala: Department of Finance, CTA, June 2005.

Painter, J. (2006). Prosaic geographies of stateness. *Political Geography* 25(7), 752–774.

Painter, J. (2007). Stateness in action. *Geoforum* 38(4), 605–607.

Painter, J. (2010). Rethinking territory. *Antipode* 42(5), 1090–1118.

Painter, J. & Jeffrey, A. (2009). *Political Geography*, 2nd edn. London: Sage.

Palakshappa, T.C. (1978). *Tibetans in India: A Case Study of Mundgod Tibetans*. New Delhi: Sterling Publishers.

Pegg, S.M. (1998). *International Society and the De Facto State*. Aldershot: Ashgate.

People's Republic of China (1992). Tibet – its ownership and human rights situation. PRC Government White Paper, September 1992. Available from www.china.org.cn/e-white/tibet/9-1.htm (accessed 14 September 2015).

Petech, L. (1973). *Aristocracy and Government in Tibet, 1728–1959*. Rome: Istituto Italiano per il medio ed Estremo Oriente.

Philo, C. & Parr, H. (2000). Editorial: Institutional geographies: introductory remarks. *Geoforum* 31(4), 513–521.

Phuntsog, J. (2005). 6.4b yuan worth of projects approved in Tibet. *South China Morning Post*, 31 May.

Pickerill, J. & Chatterton, P. (2006). Notes towards autonomous geographies. Creation, resistance and self-management as survival tactics. *Progress in Human Geography* 30(6), 1–17.

Pierson, C. (1996). *The Modern State*. London: Routledge.

Pile, S. (1997). Introduction: opposition, political identities and spaces of resistance. In S. Pile & M. Keith (eds), *Geographies of Resistance*. London: Routledge.

Pile, S. & Thrift, N. (1995). Introduction. In S. Pile and N. Thrift (eds), *Mapping the Subject: Geographies of Cultural Transformation*. London: Routledge, pp. 1–12.

Pirie, F. (2005). Segmentation within the state: the reconfiguration of Tibetan tribes in China's reform period. *Nomadic Peoples* 9(1-2), 83–102.

Pitt-Rivers, J.A. (1968). The stranger, the guest and the hostile host: introduction to the study of the laws of hospitality. In J.-G. Peristiany (ed.), *Contributions to Mediterranean Sociology: Mediterranean Rural Communities and Social Change*. Paris: Mouton, pp. 13–30.

Planning Council (1994). *Integrated Development Plan II: 1995–2000*. Dharamsala: Planning Council, CTA.

Planning Council (2000). *Tibetan Demographic Survey*. Dharamsala: Planning Council, Central Tibetan Administration.

Planning Commission (2004). *Tibetan Community in Exile: Demographic and Socio-Economic Issues, 1998–2001*. Dharamsala: Central Tibetan Administration.

Planning Commission (2010). *Demographic Survey of Tibetans in Exile – 2009*. Dharamsala: Central Tibetan Administration.

Pollard, J., McEwan, C. & Hughes, A. (2012). *Postcolonial Economies*. London: Zed.

Poulantzas, N. (1978). *State, Power, Socialism*. London: New Left Books.

Powers, J. (2004). *History as Propaganda: Tibetan Exiles versus the People's Republic of China*. Oxford: Oxford University Press.

Pratt, G. (2009). Performance. In D. Gregory, R.J. Johnston, G. Pratt & S. Whatmore (eds), *The Dictionary of Human Geography*, 5th edn. Oxford: Blackwell, pp. 525–526.

Prost, A. (2006). The problem with 'rich refugees' sponsorship, capital, and the informal economy of Tibetan refugees. *Modern Asian Studies* 40(1), 233–253.

Rabgey, T. & Wangchuk, T. (2004). Sino-Tibetan dialogue in the post-Mao era: lessons and prospects. *Policy Studies 12*. Washington, DC: East-West Center Washington.

Rajaram, P.K. & Soguk, N. (2006). Introduction: geography and the reconceptualisation of politics. *Alternatives* 31(4), 367–376.

Ramadan, A. (2008). The guests' guests: Palestinian refugees, Lebanese civilians, and the war of 2006. *Antipode* 40(4), 658–677.

Ramadan, A. (2009). A refugee landscape: writing Palestinian nationalisms in Lebanon. *ACME: An International E-Journal for Critical Geographies* 8(1): 69–99.

Ramadan, A. (2012). Spatialising the refugee camp. *Transactions of the Institute of British Geographers* 38(1), 65–77.

Regalsky, P. & Laurie, N. (2007). 'The school, whose place is this'? The deep structures of the hidden curriculum in indigenous education in Bolivia. *Comparative Education* 43(2), 231–351.

Reid-Henry, S. (2007). Exceptional sovereignty? Guantanamo Bay and the recolonial present. *Antipode* 39(4), 627–648.

Reisman, M. (1991). Governments-in-exile: Notes towards a theory of formation and operation. In Y. Shain (ed.), *Governments-in-exile in Contemporary World Politics*. London: Routledge, pp. 238–248.

Relyea, S. (1998). Trans-state entities: postmodern cracks in the great Westphalian dam. *Geopolitics* 3(2), 30–61.

Riles, A. (ed.) (2006). *Documents: Artifacts of Modern Knowledge*. Ann Arbor: University of Michigan Press.

Robinson, E.H. (2013). The distinction between state and government. *Geography Compass* (7/8), 556–566.

Robinson, J. (2003). Postcolonialising geography: tactics and pitfalls. *Singapore Journal of Tropical Geography* 24(3), 273–289.

Roemer, S. (2008). *The Tibetan Government-in-Exile: Politics at Large*. Oxford: Routledge.

Rose, N. & Miller, P. (1992). Political power beyond the state: problematics of government. *The British Journal of Sociology* 43(2), 173–205.

Rose, P.I. (ed.) (2005). *The Dispossessed: An Anatomy of Exile*. Amherst, MA: University of Massachusetts Press.

Rosello, M. (2001). *Postcolonial Hospitality*. Stanford, CA: Stanford University Press.

Rose-Redwood, R.S. (2006). Governmentality, geography and the geo-coded world. *Progress in Human Geography* 30(4), 469–486.

Roslington, J. (2013). Unrecognised states in the interwar world. Paper presented at the Producing Legitimacy: Governance against the Odds, University of Cambridge, 22–23 April 2013.

Rougier, B. (2007). *Everyday Jihad: The Rise of Militant Islam among Palestinians in Lebanon* (P. Ghazaleh, trans.). Cambridge, MA: Harvard University Press.

Routledge, P. (1996). Third space as critical engagement. *Antipode* 28(4), 399–419.

Routledge, P. (2003). Voices of the dammed: discursive resistance amidst erasure in the Narmada Valley. *Political Geography* 22(3), 243–270.

Rubin, B.R. (2007). Saving Afghanistan. *Foreign Affairs* 86(1), 57–74, 76–78.

Ruddick, S. (2004). Activist geographies: building possible worlds. In P. Cloke, P. Crang & M. Goodwin (eds), *Envisioning Human Geographies*. London: Arnold, pp. 229–241.

Ruggie, J. (1993). Territoriality and beyond: problematising modernity in international relations. *International Organisation* 47(1), 139–174.

Sack, R.D. (1986). *Human Territoriality: Its Theory and History*. Cambridge, UK: Cambridge University Press.

Said, E. (1984). The mind of winter: reflections on life in exile. *Harper's* 269(1612), 49–55.

Said, E. (2001). *Reflections on Exile and Other Literary and Cultural Essays*. London: Granta Books.

Samaddar, R. (2003). Power and care: building the new Indian state. In R. Samaddar (ed.), *Refugees and the State: Practices of Asylum and Care in India*. New Delhi: Sage, pp. 21–68.

Samuel, G. (1982). Tibet as a stateless society and some Islamic parallels. *The Journal of Asian Studies* 41(2), 215–229.

Samuel, G. (1993). *Civilized Shamans: Buddhism in Tibetan Societies*. Washington, DC: Smithsonian Institute Press.

San Martin, P. (2010). *Western Sahara: The Refugee Nation*. Chicago: University of Chicago Press.

Sangay, L. (2003). Tibet: exiles' journey. *Journal of Democracy* 14(3), 119–130.

Schechner, R. (1987). Preface: Victor Turner's last adventure. In V. Turner (ed.), *The Anthropology of Performance*. New York: PAJ Publications, pp. 7–20.

Schechner, R. (2002). *Performance Studies: An Introduction*. London: Routledge.

Schwartz, R. (1999). Renewal and resistance: Tibetan Buddhism in the modern era. In I. Harris (ed.), *Buddhism and Politics in Twentieth-Century Asia*. London: Pinter, pp. 229–253).

Scott, J.C. (1998). *Seeing Like a State: How Certain Schemes to Improve the Human Condition have Failed*. New Haven, CT: Yale University Press.

Scott, J.C. (2009). *The Art of Not Being Governed: An Anarchist History of Upland Southeast Asia*. New Haven, CT: Yale University Press.

Secor, A.J. (2001b). Ideologies in crisis: political cleavages and electoral politics in Turkey in the 1990s. *Political Geography* 20(5), 539–560.

Sennett, R. (1980). *Authority*. New York: W.W. Norton.

Shain, Y. (1989). *The Frontier of Loyalty: Political Exiles in the Age of the Nation-State*. Ann Arbor, MI: University of Michigan Press.

Shain, Y. (ed.) (1991). *Governments-in-exile in Contemporary World Politics*. London: Routledge.

Shakabpa, T.W.D. (1984). *Tibet: A Political History*. New York: Potala Publications.

Shakabpa, T.W.D. (2010). *One Hundred Thousand Moons: An Advanced Political History of Tibet, Volume 1* (D.F. Maher, trans.). Leiden: Brill.

Shakya, T. (1999). *The Dragon in the Land of Snows: A History of Modern Tibet Since 1947*. New York: Columbia University Press.

Shapiro, M.J. (1996). Introduction. In M.J. Shapiro & H.R. Alker (eds), *Challenging Boundaries: Global Flows, Territorial Identities*. Minneapolis, MN: University of Minnesota Press, pp. i–xxii.

Sharp, J. (2013). Geopolitics at the margins? Reconsidering genealogies of critical geopolitics. *Political Geography* 37, 20–29.

Shelley, T. (2004). *Endgame in the Western Sahara: What Future for Africa's Last Colony?* London: Zed Books.

Shils, E. (1965). Charisma, order, and status. *American Sociological Review* 30(2), 199–213.

Shimazu, N. (2014). Diplomacy as theatre: staging the Bandung Conference of 1955. *Modern Asian Studies* 48(1), 225–252.

Shiromany, A.A. (ed.) (1998). *The Political Philosophy of HH the XIV Dalai Lama: Selected Speeches and Writings*. New Delhi: TPPRC.

Shneiderman, S. (2006). Barbarians at the border and civilizing projects: analyzing ethnic and national identities in the Tibetan context. In P.C. Klieger (ed.), *Tibetan Borderlands*. Leiden: Brill, pp. 9–34.

Shneiderman, S. (2010). Are the Central Himalayas in Zomia? Some scholarly and political considerations across time and space. *Journal of Global History* 5, 289–312.

Shore, C. & Wright, S. (eds) (1997). *Anthropology of Policy: Critical Perspectives on Governance and Power*. London: Routledge.

Shryock, A. (2012). Breaking hospitality apart: bad hosts, bad guests, and the problem of sovereignty. *Journal of the Royal Anthropological Institute* 18(S1), S20–S33.

Sidaway, J.D. (2002). *Imagined Regional Communities: Integration and Sovereignty in the Global South*. London: Routledge.

Sidaway, J.D. (2003). Sovereign excesses? Portraying postcolonial sovereigntyscapes. *Political Geography* 22(2), 157–178.

Siddiqui, A. (1991). Muslims of Tibet. *Tibet Journal* 16(4), 71–85.

Sidhu, R. & Christie, P. (2007). Spatialising the scholarly imagination: globalisation, refugees and education. *Transnational Curriculum Inquiry* 4(1), 7–17.

Sloane, R.D. (2014). Tibetan Diaspora in the shadow of the self-immolation crisis: consequences of colonialism. In S. Akram & T. Syring (eds), *Still Waiting for Tomorrow: the Law and Politics of Unresolved Refugee Crises*. Newcastle upon Tyne: Cambridge Scholars Publishing, pp. 55–74.

Smith, A.D. (1981). States and homelands: the social and geopolitical implications of national territory. *Millennium: Journal of International Studies* 10(3), 187–201.

Smith, W.W. (2008). *China's Tibet? Autonomy or Assimilation*. Plymouth, UK: Rowman & Littlefield.

Sneath, D. (2007). *The Headless State: Aristocratic Orders, Kinship Society and Misrepresentations of Nomadic Inner Asia*. New York: Columbia University Press.

Soguk, N. (1996). Transnational/transborder bodies: resistance, accommodation and exile in refugee and migration movements on the US-Mexico Border. In M.J. Shapiro & H.R. Alker (eds), *Challenging Boundaries: Global Flows, Territorial Identities*. Minneapolis: University of Minnesota Press, pp. 285–326.

Soguk, N. & Whitehall, G. (1999). Wandering grounds: transversality, identity, territoriality and movement. *Millennium: Journal of International Studies* 28(3), 675–698.

Sonam, B.D. (ed.) (2004). *Muses in Exile: An Anthology of Tibetan Poetry*. New Delhi: Paljor Publications.

Southall, A. (1956). *Alur Society: A Study in Processes and Types of Domination*. Cambridge: W. Heffer & Sons.

Sparke, M. (2013). From global dispossession to local repossession: towards a worldly cultural geography of Occupy activism. In N.C. Johnson, R.H. Schein & J. Winders (eds), *The Wiley-Blackwell Companion to Cultural Geography*. Oxford: Wiley-Blackwell, pp. 387–408.

Spears, I.S. (2004). States-within-states: an introduction to their empirical attributes. In P. Kingston & I.S. Spears (eds), *States-Within-States: Incipient Political Entities in the Post Cold War Era*. New York: Palgrave Macmillan, pp. 15–34.

Sperling, E. (2001). "Orientalism" and aspects of violence in the Tibetan tradition. In T. Dodin & H. Räther (eds), *Imagining Tibet: Perceptions, Projections, and Fantasies*. Somerville, MA: Wisdom, pp. 315–329.

Sperling, E. (2004). The Tibet-China conflict: history and polemics. *Policy Studies 7*. Washington DC: East-West Center Washington.

Starkweather, S. (2009). Governmentality, territory and the U.S. census: The 2004 Overseas Enumeration Test. *Political Geography* 28, 239–247.

Steinberg, P.E. & Chapman, T.E. (2009). Key West's Conch Republic: Building sovereignties of connection. *Political Geography* 28(5), 283–295.

Stepputat, F. (1992). Beyond relief? Life in a Guatemalan refugee settlement in Mexico. PhD dissertation. University of Copenhagen.

Sternberger, D. (1968). Legitimacy. In D.L. Sills (ed.), *International Encyclopaedia of the Social Sciences*, Vol. 9. New York: Macmillan, pp. 244–248.

Stoddard, H. (1994). Tibetan publications and national identity. In R. Barnett (ed.), *Resistance and Reform in Tibet*. Bloomington, IN: Indiana University Press, pp. 121–156.

Stoler, L.A. (2007). Affective states. In D. Nugent & J. Vincent (eds), *A Companion to the Anthropology of Politics*. Oxford: Blackwell, pp. 4–20.

Strang, D. (1996). Contested sovereignty: the social construction of colonial imperialism. In T.J. Biersteker & C. Weber (eds), *State Sovereignty as Social Construct*. Cambridge, UK: Cambridge University Press, pp. 22–49.

Strathern, M. (ed.) (2000). *Audit Cultures: Anthropological Studies in Accountability, Ethics, and the Academy*. New York: Routledge.

Strauss, M.J. (2007). Guantanamo Bay and the evolution of international leases and servitudes. *The New York City Law Review* 10(2), 479–510.

Ström, A.K. (1995). The quest for grace: identification and cultural continuity in the Tibetan Diaspora. *Oslo Occasional Papers in Social Anthropology* 24.

Strong, A. (1912). Some aspects of the Tibetan problem. *Contemporary Review* 102, 527–533.

Subba, T.B. (1990). *Flight and Adaptation: Tibetan Refugees in the Darjeeling-Sikkim Himalaya*. Dharamsala: Library of Tibetan Works and Archives.

Subramanya, N. (2004). *Human Rights and Refugees*. New Delhi: A P H Publishing Corporation.

Talmon, S. (1998). *Recognition of Governments in International Law: With particular reference to governments-in-exile*. Oxford: Clarendon Press.

Tambiah, S.J. (1977). The Galactic Polity: The structure of traditional kingdoms in Southeast Asia. *Annals of the New York Academy of Sciences* 293, 69–97.

Taussig, M. (1992). *The Nervous System*. London: Routledge.

Taussig, M. (1997). *The Magic of the State*. London: Routledge.

Taylor, D. (1995). Performing gender: Las Madres de la Plaza de Mayo. In D. Taylor & J. Villegas (eds), *Negotiating Performance: Gender, Sexuality, and Theatricality in Latin/o America*. Durham, NC: Duke University Press, pp. 275–305.

Taylor, P.J. (1994). The state as container: territoriality in the modern world-system. *Progress in Human Geography* 18(2), 151–162.

TCC (2006). *Tibetan Chamber of Commerce: Report of Proceedings of The First Annual General Meeting November 23–24, 2006, Dharamsala*. New Delhi: Tibetan Chamber of Commerce.

TechnoServe (2010). *The economic development program of Tibetan refugees in India: Improving the quality of life through the creation and enhancement of livelihood opportunities*. Mumbai: TechnoServe.

Tenzin Dorjee (2013). Diplomacy or mobilization? The Tibetan dilemma in the struggle with China. In J.-H. Bae & J.H. Ku (eds), *China's Internal and External Relations and Lessons for Korea and Asia*. Seoul: Korea Institute for National Unification, pp. 63–112.

Tenzin Sherab. (2012). A review of the Central Tibetan Administration's poverty programme. *Tibetan Review* 47(7–9), 22.

TGiE (1992). *Guidelines for Future Tibet's Polity and basic features of its Constitution* (draft translation). Dharamsala, Central Tibetan Administration.

The Economist (2001). Home thoughts from abroad. *The Economist* 22 December, pp. 45–46.

The Economist (2008). Tibet: Britain's suzerain remedy. *The Economist* 6 November, p. 64.

The Tibet Post (2010). Tibetans Urged to Participate in 2011 Indian National Census. *The Tibet Post International*, 13 May. Available from www.thetibetpost.com/en/outlook/interviews-and-recap/866-tibetans-urged-to-participate-in-2011-indian-national-census (accessed 15 September 2015).

Thomassen, B. (2012). Notes towards an anthropology of political revolutions. *Comparative Studies in Society and History* 54(3), 679–706.

Thrift, N. (2000). It's the little things. In K. Dodds & D. Atkins (eds), *Geopolitical Traditions: A Century of Geopolitical Thought*. London: Routledge, pp. 380–387.

Thupten Samphel (2006). Majnuka Tilla re-visited. *Tibetan Bulletin* 10(5), 27–28.

Tibet Justice Center (2011). Legal issues implicated by the Dalai Lama's devolution of power. Republished by *Tibetan Political Review*. Available from https://sites.google.com/site/tibetanpoliticalreview/articles/tibetjusticecenterreleaseslegalmemorandumondalailamashistoricdevolutionofpower (accessed 16 September 2015).

Tibet Justice Center (2012). *Tibet's Stateless Nationals II: Tibetan Refugees in India*. Oakland: TJC.

Tibet.net (2012) 8th Annual Conference of Tibetan Settlement Officers begins. Central Tibetan Administration, June 14. Available from http://tibet.net/2012/06/8th-annual-conference-of-tibetan-settlement-officers-begins/ (accessed 15 September 2015).

Tibetan Bulletin (2001). Prof. Samdhong Rinpoche speaking to the press after taking his oath of office. Lhakpa Tsering Hall, Dharamsala. 5 September 2001. *Tibetan Bulletin* 5(4), 25.

Tibetan Bulletin (2003). Exile economics. *Tibetan Bulletin* 7(3), 16.

Tibetan Political Review (2011). Sangay's new Cabinet: strengths and weaknesses. *Tibetan Political Review* 26 September. Available from https://sites.google.com/site/tibetanpoliticalreview/articles/sangay%E2%80%99snewcabinetstrengthsandweaknesses (accessed 15 September 2015).

Tibetan Political Review (2013a). To be or not be: should Tibetans in India assert Indian citizenship? *Tibetan Political Reivew* 26 April. Available from https://sites.google.com/site/tibetanpoliticalreview/editorials/tobeornotbe (accessed 15 September 2015).

Tibetan Political Review (2013b). Tibetan democracy takes a step backwards. *Tibetan Political Review* 24 October. Available from http://www.tibetanpoliticalreview.org/editorials/tibetandemocracytakesastepbackwards (accessed 15 September 2015).

Tibetan Review (2008). Exile government outlines position vis-à-vis NGOs. *Tibetan Review* 43(3), 26–27.

Tibetan Women's Association (2005). *The Status of Exiled Tibetan Women in India*. Dharamsala: Tibetan Women's Association.

Till, K. (2003). Places of memory. In J. Agnew, K. Mitchell & G. Toal (eds), *A Companion to Political Geography*. Oxford: Blackwell, pp. 289–301.

TPPRC (2003). *Tibet's Parliament in Exile*. New Delhi: Tibetan Parliamentary and Policy Research Centre.

TPPRC (2006). *Indian Parliament on the Issue of Tibet: Lok Sabha Debates 1952–2005*. New Delhi: Tibetan Parliamentary and Policy Research Centre.

Travers, A. (2011). The careers of the noble officials of the Ganden Phodrang (1895–1959): organisation and hereditary divisions within the service of state. In K.N. Gurung, T. Myatt, N. Schneider & A. Travers (eds), *Revisiting Tibetan Culture and History, Proceedings of the Second International Seminar of Young Tibetologists, Paris, 2009, Volume 1, Revue d'Études Tibétaines n° 21*. Paris: CNRS, pp. 155–174.

Travers, A. (2012). Patrimonial aspects of public service: the careers of the Ganden Phodrang officials (1895–1959). In R. Vitali (ed.), *Studies on the History and Literature of Tibet and the Himalaya*. Kathmandu: Vajra Publications, pp. 157–181.

Trouillot, M.-R. (2001). The anthropology of the state in the age of globalization: close encounters of the deceptive kind. *Current Anthropology* 42(1), 125–138.

Trouillot, M.-R. (2003). *Global Transformations: Anthropology and the Modern World*. New York: Palgrave.

Tsundue, T. (2003). *Kora: Stories and Poems*. Dharamsala: TibetWrites.

Tsundue, T. (2004). Sontsa: Tibetan youth power. *Tibetan Review* 39(8), 19.

Turner, S. (2005). Suspended spaces – contesting sovereignties in a refugee camp. In T.B. Hansen & F. Stepputat (eds), *Sovereign Bodies: Citizens, Migrants, and States in the Postcolonial World*. Princeton, NJ: Princeton University Press, pp. 312–332.

Turner, V. (1974). *Dramas, Fields, and Metaphors: Symbolic Action in Human Society*. Ithaca, NY: Cornell University Press.

Turner, V. (1982). *From Ritual to Theatre: The Human Seriousness of Play*. Baltimore: PAJ Press.

Turner, V. (1987a). *The Anthropology of Performance*. New York: PAJ Publications.

Turner, V. (1987b). Betwixt and between: the liminal period in rites of passage. In L.C. Mahdi, S. Foster & M. Little (eds), *Betwixt & Between: Patterns of Masculine and Feminine Initiation*. Peru, IL: Open Court Publishing, pp. 2–19.

Tuttle, G. (2005). *Tibetan Buddhists in the Making of Modern China*. New York: Columbia University Press.

Van Hear, N. (2003). From Durable Solutions to Transnational Relations: Home and Exile among Refugee Diasporas. *New Issues in Refugee Research. Working Paper No. 83, March 2003*. Geneva: UNHCR Evaluation and Policy Analysis Unit.

van Schendel, W. (2002). Stateless in South Asia: The making of the India-Bangladesh enclaves. *Journal of Asian Studies* 61(1), 115–147.

van Walt van Praag, M. (1987). *Status of Tibet: History, Rights and Prospects in International Law*. Boston: Wisdom Publications.

van Walt van Praag, M. (2013). A legal examination of the 1913 Mongolia–Tibet Treaty of Friendship and Alliance. *Lungta, Spring 2013: Special edition: The Centennial of the Tibeto-Mongol Treaty 1913–2013*.

Vasudevan, A. (2015). The autonomous city: Towards a critical geography of occupation. *Progress in Human Geography* 39(3), 316–337.

Vigne, R. (1987). SWAPO of Namibia: a movement in exile. *Third World Quarterly* 9(1), 85–107.

Vinokurov, E. (2007). *A Theory of Enclaves*. Lantham, MD: Lexington Books.

Vrasti, W. (2008). The strange case of ethnography and international relations. *Millennium: Journal of International Studies* 37(2), 279–301.

Wahlbeck, O. (1998). Transnationalism and diasporas: the Kurdish example. Paper presented at the International Sociological Association XIV World Congress of Sociology, Montreal, Canada.

Wakefield, S.E.L. (2007). Reflective action in the academy: exploring praxis in critical geography using a 'food movement' case study. *Antipode* 39(2), 331–354.

Walker, R.J.B. (1993). *Inside/Outside: International Relations as Political Theory*. Cambridge: Cambridge University Press.

Wangdu, K. (2013). Special Frontier Force: A clandestine Tibetan guerrilla unit in the Indian military. *Merabsarpa* 6 January. Available from http://www.merabsarpa. com/politics/special-frontier-force (accessed 15 September 2015).

Watts, M. (2003). Development and governmentality. *Singapore Journal of Tropical Geography* 24(1), 6–34.

Watts, N.F. (2004). Institutionalizing Virtual Kurdistan West. In J.S. Migdal (ed.), *Boundaries and Belonging: States and Societies in the Struggle to Shape Identities and Local Practices*. Cambridge: Cambridge University Press, pp. 121–147.

Weber, C. (1995). *Simulating Sovereignty: Intervention, the State and Symbolic Exchange*. Cambridge, UK: Cambridge University Press.

Weber, C. (1998). Performative states. *Millennium: Journal of International Studies* 27(1), 78–95.

Weber, M. (1919/2009). Politics as a vocation. In H.H. Gerth & C. Wright Mills (eds), *Max Weber: Essays in Sociology*. London: Routledge & Kegan Paul, pp. 77–128.

Weber, M. (1922/2009). *Bureaucracy*. In H.H. Gerth & C. Wright Mills (eds), *Max Weber: Essays in Sociology*. London: Routledge & Kegan Paul, pp. 196–232.

Weber, M. (1947). *The Theory of Social and Economic Organisation*. Glencoe, IL: Free Press.

Weber, M. (1978). *Economy and Society: Volume I*. New York: Bedminster.

Wei, J. (ed.) (1989). *100 Questions about Tibet*. Beijing: Beijing Review Publications.

Weizman, E. (2011). *The Least of All Possible Evils: Humanitarian Violence from Arendt to Gaza*. London: Verso.

Wilson, A. (forthcoming). *Remaking Sovereignty: State Power, Revolution and Exile in a Saharan Liberation Movement*. Philadelphia: University of Pennsylvania Press.

Wilson, A. & McConnell, F. (forthcoming). Constructing legitimacy without legality in long term exile: Comparing Western Sahara and Tibet. *Geoforum*.

White, G.W. (2004). *Nation, State, and Territory*. Lanham, MD: Rowman & Littlefield.

Wunderlich, F.M. (2010). The aesthetics of place-temporality in everyday urban space: the case of Fitzroy Square. In T. Edensor (ed.), *Geographies of Rhythm: Nature, Place, Mobilities and Bodies*. Aldershot: Ashgate, pp. 45–56.

Yamamoto, L. & Esteban, M. (2010). Vanishing island states and sovereignty. *Ocean and Coastal Management* 53, 1–9.

Yeh, E. (2002). Will the real Tibetan please stand up! Identity politics in the Tibetan diaspora. In P.C. Klieger (ed.), *Tibet, Self, and the Tibetan Diaspora: Voices of Difference*. Leiden: Brill, pp. 229–254.

Yeh, E.T. (2007). Exile meets homeland: politics, performance and authenticity in the Tibetan diaspora. *Environment and Planning D: Society and Space* 25(4), 648–667.

Yeh, E.T. (2013). *Taming Tibet: Landscape Transformation and the Gift of Chinese Development*. Ithaca: Cornell University Press.

Yeh, E.T. & Lama, K.T. (2006). Hip-hop gangsta or most deserving victims? Transnational migrant identities and the paradox of Tibetan racialization in the USA. *Environment and Planning A* 38(5), 809–829.

Žižek, S. (1995). *The Sublime Object of Ideology*. London: Verso.

Index

Rehearsing the State: The Political Practices of the Tibetan Government-in-Exile, First Edition. Fiona McConnell.
© 2016 John Wiley & Sons, Ltd. Published 2016 by John Wiley & Sons, Ltd.